PRECISE
NUMERICAL METHODS
USING C++

PRECISE NUMERICAL METHODS USING C++

Oliver Aberth

Mathematics Department
Texas A & M University

ACADEMIC PRESS

San Diego London Boston New York Sydney Tokyo Toronto

This book is printed on acid free paper. ∞

Copyright © 1998 by Academic Press

ACADEMIC PRESS
525 B Street, Suite 1900, San Diego, CA 92101-4495, USA
1300 Boylston Street, Chestnut Hill, MA 02167, USA
http://www.apnet.com

United Kingdom Edition published by
ACADEMIC PRESS LIMITED
24-28 Oval Road, London NW1 7DX
http://www.hbuk.co.uk/ap

Library of Congress Cataloging-in-Publication Data

Library of Congress CIP data pending.

ISBN 0-12-041750-2

Printed in the United States of America
98 99 00 01 02 IP 10 9 8 7 6 5 4 3 2 1

to Dawn

CONTENTS

III

SOLVABLE PROBLEMS AND NONSOLVABLE PROBLEMS

IV

COMPUTING DERIVATIVES AND INTEGRALS

V

FINDING ZEROS OF REAL FUNCTIONS

VI

FINDING ZEROS OF POLYNOMIALS AND OTHER ANALYTIC FUNCTIONS

VII

PROBLEMS OF LINEAR ALGEBRA

VIII

OPTIMIZATION PROBLEMS

IX

NUMERICAL SOLUTION OF ORDINARY DIFFERENTIAL EQUATIONS

X

THE C++ SYSTEM FOR PRECISE COMPUTATION

ACKNOWLEDGMENTS

In the writing of this book and in the creation of the accompanying software, I have had help from many sources.

Two people have made central contributions. Mark J. Schaefer, formerly of Tübingen University, helped write the initial C++ code. His brilliant programming skills were much needed, and it was his idea to identify quantities correct to the last decimal place with a terminal tilde. Ramon Moore of Ohio State University has been supportive through the last decade, and he helped test the various computation programs. The many cases of not so stellar performance he has pointed out have led to the steady improvement of the routines. Professor Moore also looked at a preliminary version of this book and made many helpful suggestions and corrections.

The National Science Foundation supported the development of the range arithmetic system over the period from 1988 to 1996.

Norman Naugle of Texas A & M University supplied the TeX software used in writing this book. The diagrams were formed using the PicTeX program of Michael J. Wichura of the University of Chicago. Steve Johnson, my department's computing system administrator, solved many technical glitches.

I am indebted to the interval arithmetic research community for help with the demonstration programs. Rudolf Lohner of Karlsruhe University made me aware of the recent developments in the interval treatment of differential equations. R. Baker Kearfott of the University of Southwestern Louisiana gave similar guidance with respect to topological degree methods. Many other researchers pointed out a deficiency in a specific program or provided some idea that was used to correct a difficulty.

My thanks to all, many unmentioned, for their help.

Oliver Aberth

PREFACE

To give meaning to the adjective *precise* in the book title, first consider the conventional way of numerically solving a mathematical problem: An answer is computed by some method, and then possibly the answer's error is estimated. With this approach there are various ways to run into difficulty. The simplest way is when we are unsuccessful in obtaining an answer, the problem being too large or too difficult for our computing resources. But even when an answer is obtained, if we have only a vague idea of the answer's error, there are two distinct ways we could be in trouble. It could be that the answer is right on the mark, but because we are uncertain of the answer's accuracy, we refrain from using the answer to the full extent it deserves. More seriously, the answer could be completely spurious, but we make use of it anyway, not having any indication of the huge error.

In this book, the *precise* numerical solution of a mathematical problem means that a specific number of correct decimal places for the answer is prescribed in advance, and the computer then does whatever is necessary to achieve this result. Here failure can occur in exactly one way: No answer is obtained, the problem being too large or too difficult for our computing resources. We never encounter either of the other two difficulties of the conventional approach.

It is clear that the precise solution of a mathematical problem takes more computing horsepower than the conventional approach, because not only must the computer generate an answer, but it must keep track of the errors it is making in its computation. With the computers available fifteen or twenty years ago, it would have been unrealistic to attempt this precise approach. But today the Macintoshes and PCs we use are awesome in their power, and multiple processor computers also are becoming available. The CD-ROM included with this book has a suite of demonstration programs that most PCs can run. These demonstration programs show the ease with which the typical problems of elementary numerical analysis can be solved precisely. The C++ text files for these demonstration programs are available via the Web and, with the appropriate C++ compiler, can be converted into executable programs for any computer.

The *Exercises* sections suggest test problems that the reader can solve with the demonstration programs, which help to illustrate the points brought up in the text. The *Notes* sections give references to sources for the methods presented.

This text can be used for a course in numerical analysis for advanced undergraduate or graduate students. It assumes the reader is familiar with linear algebra and with differential equations. Chapter 6 requires some understanding of the theory of complex analytic functions.

1

INTRODUCTION

1.1. PRECISE NUMERICAL ANALYSIS

Precise numerical analysis may be defined as the study of computer methods for solving mathematical problems either exactly or to prescribed accuracy. Specifically, we require the following: If the solution of a mathematical problem requires finding certain real numbers, then these numbers must be found to k correct decimal places, where the integer k is specified in advance. Likewise, if certain complex numbers are required, then both real and imaginary parts of these numbers must be found to k correct decimal places. For certain problems with rational number solutions, we consider also whether it is possible to obtain each rational number exactly in the form p/q, where p and q are integers. But these problems are a small minority of the problems treated, and the usual objective will be to obtain our answers to a specified number of correct decimal places.

When attempting the precise solution of a given mathematical problem, the first step must always be the analysis of the problem, to see whether *in principle* it can always be solved with a computer. That is, the solution of the problem must have no *intrinsic* computational difficulties. Such a problem will be called a *solvable* problem, whereas a problem that fails this analysis is a *nonsolvable* problem. In Chapter 3 we give the mathematical basis for this approach.

After we are certain our mathematical problem is solvable, we then can consider practical methods of solution. Solving a mathematical problem to prescribed accuracy generally requires more computation than solving it with ordinary numerical methods, because besides getting answers, we must bound our errors. Here we will introduce *interval arithmetic* to aid us with the computation of error bounds. Interval arithmetic was introduced in 1962 by the American mathematician Ramon Moore, and now it has developed into a mature subject through his efforts and those of many mathematicians around the world. Ordinary floating-point arithmetic, the usual arithmetic of computation, is inadequate for our needs because no error indication is obtained, and this arithmetic must be improved in two ways before it can be used. First, the precision of computation needs to be changeable during computation. In simple terms, if we are carrying a certain number of decimal digits in our computation, and it turns out that our answers are not accurate enough, then we must repeat the computation using more digits. However, to decide whether we are carrying enough decimal digits, we must determine which digits of a computed result are reliable and which digits are not. This leads to the second improvement of adding a small *range* field to each floating-point number and using interval arithmetic. Floating-point arithmetic with these two extensions is *range arithmetic*, and we describe it in the next chapter. Also in the next chapter, we describe a system for rational arithmetic, which is used for the occasional problem requiring rational answers.

Following these preliminary chapters, each of Chapters 4 through 9 deals with a particular problem area. These chapters list the solvable problems and describe practical algorithms for solving them. Chapter 10 has a description of our C++ system and a list of the demonstration programs. The files for the C++ system and for the demonstration programs can be obtained electronically; the procedure is given in Chapter 10.

1.2. THE COSTS AND BENEFITS OF PRECISE COMPUTATION

The costs are evident: for the computer, more extensive calculations, and for the programmer, attention to the analysis and computation of error. In each program, at the point where the final answers are obtained, these answers require checking that the desired number of correct decimal places have been obtained. If not, then the number of digits carried in the computation is increased and the entire computation is redone. As our computers get ever more powerful and as multiple processor computers become available, the extra computation becomes unimportant for simple problems, typically for all the sample problems treated in an elementary numerical analysis text.

One small benefit is that many programming errors that would go undetected for an ordinary computational approach are caught when precise computation is done. Clearly a gross error makes itself felt directly and is corrected in either approach. But a subtle error that only leads to a small deviation in an answer may not be noticed with the routine approach. With precise computation, whenever a test problem is solved, any deviation in the computed answer from the true value, however small, signals a programming error.

Another minor benefit is that a user of a precise computation program has less need to know the details of the computation. With a conventional program, the user should know the characteristics of the solution procedure so that grossly inaccurate results can be recognized and rejected. With a precise computation program, the possibility of obtaining incorrect results for a *properly posed problem* is nil, assuming, of course, that all programming blunders have been caught and corrected. There still is a narrow possibility of obtaining a numerical answer to a problem that actually has no solution, the problem not being properly posed (see Section 3.5), but this danger is relatively minor.

A major benefit is that when a precise solution program fails to solve a problem, this always happens in the same safe way: the computer memory is exhausted or the program execution time is too great. In this book we call such problems *computerbound* problems. A failure this way is less troublesome than dealing with erroneous results that we may accept unknowingly when a conventional solution program fails. Obtaining a solution to a computerbound problem requires getting around the blockage. If more computer memory is needed, then this must be obtained. If the difficulty is execution time, perhaps we can let the computer run overnight. Or we can obtain a more powerful machine. Or, if these options are not practical, it may be possible to develop a specialized precise computation program for our particular problem.

Perhaps the most significant benefit is that experimental results can be better utilized with precise computation. Usually experimental results are hard to come by and, accordingly, are precious. If we try to see whether these results match a mathematical model, and our computations for the model are not precise, then we may attribute a slight descrepancy between the model and the experiment to our computational error. If we do this, however, we miss our chance to refine the model and to begin the next cycle of model improvement. With precise computation discrepancies are not misunderstood in this way.

1.3. SPECIFICATION OF CORRECT DECIMAL PLACES

The *signed error* of an approximate answer is defined by the equation

$$\text{approximate value} = \text{true value} + \text{signed error}$$

Thus the signed error may be zero, positive, or negative. Most often we are interested only in the absolute value of the signed error, and this quantity we call simply the *error*. That is,

$$\text{error} = |\text{signed error}|$$

and so the error of an approximation is always positive or zero.

Now imagine that we desire our answers to be correct to four decimal places. The distance on the real number line between any pair of neighboring four decimal-place values, say 52.7777 and 52.7778, is $0.0001 = 10^{-4}$, so each four-place number is the best four place approximation for a true value that is less than half this distance away. Whatever the true value may be, there is only one correct four decimal-place answer, except in the unusual case where the true value is situated exactly between two four decimal-place values, for instance, if the true value were 52.77775, situated midway between 52.7777 and 52.7778. In such a case two different correct four decimal-place approximation can be used, neither being superior to the other. Accordingly, it is reasonable to accept a particular four decimal-place value as the correct one whenever its error is not greater than $\frac{1}{2} \cdot 10^{-4} = 0.00005$. In general then, we take a k decimal-place approximant to be correct if

$$\text{error} \leq \frac{1}{2} \cdot 10^{-k} \tag{1.1}$$

When the true value we are approximating is either very large or very small, instead of using k decimal-place values in "fixed decimal-point" notation, as just shown, it is usually more convenient to use k decimal-place values in "scientific floating-point" notation. These are numbers of the form

$$(\text{sign})\ d_0.d_1d_2 \cdots d_k \cdot 10^e \tag{1.2}$$

with the requirement that the exponent e be an integer and that the decimal digit d_0 not be zero. For instance, $-3.4444 \cdot 10^{12}$ is a four decimal-place, scientific floating-point value, but $-3.4444 \cdot 10^{\sqrt{2}}$ or $0.4444 \cdot 10^{12}$ are not. The leading term "(sign) $d_0.d_1d_2 \ldots d_k$" is the *mantissa* part, and the trailing factor "10^e" is the *exponent* part of the number.

We follow the form of the definition of a correct k place, fixed decimal-point approximant given above, and we take a k decimal-place, scientific floating-point approximant to be correct if

$$\text{error} \leq \frac{1}{2} \cdot 10^{-k} \cdot 10^e \tag{1.3}$$

The great utility of using fixed-point and floating-point values that are correct to a certain number of decimal places is that this automatically gives us an upper bound on the error of our approximation. It is convenient to identify any number correct to the last decimal place by adding a tilde (˜) after the last decimal: 1.234˜, for example. This makes it easy to distinguish between approximations and exact rationals without the tilde: 1.2500, for example. To determine the maximum error of a displayed approximation, say 2.22˜ or the floating point number $1.1111˜ \cdot 10^{11}$, one can mentally change the tilde to a 5 and convert all preceding digits to 0s. Thus the maximum errors of these last two numbers are 0.005 and $0.00005 \cdot 10^{11}$, respectively. If the displayed decimal number is A, and ϵ is the error bound, then the true value lies somewhere in the interval $[A - \epsilon, A + \epsilon]$. We will call this interval the *tilde-interval* of the approximation.

VARIOUS ARITHMETICS

2.1. FLOATING-POINT ARITHMETIC

The earliest (one of a kind) computers provided only some form of fixed-point arithmetic. By the middle 1950s commercial computers were providing floating-point arithmetic for general scientific computation. In this system, as commonly used today, each floating-point number is assigned one fixed block of computer memory from several allowed sizes, and the programmer chooses the size. Often today there are either two or three sizes allowed. The number stored in memory has the form

$$(\text{sign}) \ 0.d_1 d_2 \cdots d_k \cdot B^e$$

where $d_1 d_2 \cdots d_k$ is the mantissa, e is the exponent, and B is the base used for the exponent part. Here various number systems can be used in this representation: binary, octal, decimal, or hexadecimal. Accordingly, the base B could be 2, 8, 10, or 16. In our C++ system we use the decimal system, so we will concentrate on decimal floating-point representations, which have the form

$$(\text{sign}) \ 0.d_1 d_2 \cdots d_k \cdot 10^e \tag{2.1}$$

Thus the representation is similar to the scientific floating-point representation mentioned in Chapter 1, except that the decimal point precedes the first digit of the mantissa instead of being after the first digit. This convention makes multiplication and division a little easier to execute on the computer. A certain amount of the assigned memory space is for the exponent e. Let us suppose that enough bits are allocated so that an exponent can vary between plus or minus 10,000, because from computational experience this is about as much variation as is ever needed. The mantissa digits take up most of the memory space, with the number of digits, k, depending on the floating-point size we choose. For example, we may have three choices available, giving us 6, 14, or 25 mantissa digits. When two numbers of a particular size are added, subtracted, or multiplied, the mantissa of an exact result may be longer than the operand mantissas. Digits then must be discarded to make the result fit into the memory block reserved for the answer.

This floating-point arithmetic is obtained either by programming the computer to do it (a software floating point), or having it done by a silicon chip (a hardware floating point). The hardware floating point is preferable, of course, because it executes much faster. A drawback of this conventional system is that the size must be chosen before a program is run. This precludes having the program choose the size after doing a sample computation. Let us pass then to a more flexible floating-point system.

2.2. VARIABLE PRECISION FLOATING-POINT ARITHMETIC

With this arithmetic the format of a number, shown below,

$$\text{(sign) } 0.d_1d_2 \cdots d_n \cdot 10^e \tag{2.2}$$

is almost identical to that given previously except that n, the number of digits in the mantissa, now is variable instead of being fixed at the few choices previously allowed. Such a system has an upper bound, N, on the number of mantissa digits that can be used, with N typically being several thousand. Thus any number of the form above is allowed if n is in the range $1 \leq n \leq N$. Here because n is not fixed, the amount of memory needed for a number is not known in advance of forming the number. The storage of a number in memory now requires a variable number of adjoining bytes to hold the mantissa. The integer n, stored in a designated initial part of a number's memory block, is used to determine the number of adjoining memory bytes holding mantissa digits.

This system is now able to correctly represent even a very long input decimal constant. We simply convert the number to the form (2.2) with n just large enough to represent all digits. Thus program constants, such as 2.75 or 2222211111, obtain the forms $+0.275 \cdot 10^1$ or $+0.2222211111 \cdot 10^{10}$, respectively. When performing divisions with such numbers, we must decide how many mantissa digits to form. When doing additions, subtractions, or multiplications, generating an exact result may require an inconveniently large number of mantissa digits.

A control parameter is needed, bounding the length of all mantissas formed by arithmetic operations. This control parameter, always a positive integer, can be given the name PRECISION. Thus PRECISION can be set to any value between a default minimum and the system upper bound N.

Before any computation, we set PRECISION to some value that seems appropriate, and if the computation reveals that more mantissa digits are needed, then we increase PRECISION and repeat the computation. Note that PRECISION does not determine the mantissa length for the decimal constants used by a program. Such constants always are converted to floating-point form with a mantissa just long enough to represent them exactly.

This system is clearly an improvement over the previous floating-point system, but there still is a major drawback. We usually cannot tell from our computed numbers how accurate our results are, especially after a long and complicated computation, so we do not know when to increase PRECISION. We need some way of determining how big our computational error is.

2.3. INTERVAL ARITHMETIC

In 1962 Ramon Moore proposed a way for the computer to keep track of errors. Instead of using a single computer representation for a number, two would be used, one to represent the number, and another to represent the maximum error of the number. Each number then has a representation $m \pm \epsilon$, where m is the computer representation of the number, just as before, and ϵ is the computer representation for the error. Program constants that are exactly represented by a computer number have ϵ set to zero. As the computer does each arithmetical operation of addition, subtraction, multiplication, and division, besides forming the result m, it forms the result's error bound ϵ. The mathematical relations used for the four arithmetic operations are the following:

$$m_1 \pm \epsilon_1 \quad + \quad m_2 \pm \epsilon_2 = (m_1 + m_2) \pm (\epsilon_1 + \epsilon_2) \tag{2.3}$$

$$m_1 \pm \epsilon_1 \quad - \quad m_2 \pm \epsilon_2 = (m_1 - m_2) \pm (\epsilon_1 + \epsilon_2) \tag{2.4}$$

$$m_1 \pm \epsilon_1 \quad \times \quad m_2 \pm \epsilon_2 = (m_1 m_2) \pm (\epsilon_1 |m_2| + |m_1| \epsilon_2 + \epsilon_1 \epsilon_2) \tag{2.5}$$

$$m_1 \pm \epsilon_1 \quad \div \quad m_2 \pm \epsilon_2 = \begin{cases} \dfrac{m_1}{m_2} \pm \left(\dfrac{\epsilon_1 + \left|\dfrac{m_1}{m_2}\right| \epsilon_2}{|m_2| - \epsilon_2} \right) & \text{if } |m_2| > \epsilon_2 \\ \text{Division error} & \text{if } |m_2| \leq \epsilon_2 \end{cases} \tag{2.6}$$

The first three equations above are clear. To obtain the error bound of the last relation, first replace m_1 by $|m_1|$ and m_2 by $|m_2|$; the error bound obtained for this case applies to the original formulation. We see that the maximum error occurs either when the smallest possible numerator is divided by the largest possible denominator or when the largest possible numerator is divided by the smallest possible denominator. In the first case the error is

$$\frac{|m_1|}{|m_2|} - \frac{|m_1| - \epsilon_1}{|m_2| + \epsilon_2} = \frac{|m_1|(|m_2| + \epsilon_2) - (|m_1| - \epsilon_1)|m_2|}{|m_2|(|m_2| + \epsilon_2)}$$

$$= \frac{\epsilon_1|m_2| + |m_1|\epsilon_2}{|m_2|(|m_2| + \epsilon_2)}$$

$$= \frac{\epsilon_1 + \left|\dfrac{m_1}{m_2}\right|\epsilon_2}{|m_2| + \epsilon_2}$$

In the other case the error is

$$\frac{|m_1| + \epsilon_1}{|m_2| - \epsilon_2} - \frac{|m_1|}{|m_2|} = \frac{(|m_1| + \epsilon_1)|m_2| - |m_1|(|m_2| - \epsilon_2)}{|m_2|(|m_2| - \epsilon_2)}$$

$$= \frac{\epsilon_1|m_2| + |m_1|\epsilon_2}{|m_2|(|m_2| - \epsilon_2)}$$

$$= \frac{\epsilon_1 + \left|\dfrac{m_1}{m_2}\right|\epsilon_2}{|m_2| - \epsilon_2}$$

If ϵ_2 is positive, this second value for the error is greater and is the error shown in (2.6).

With interval arithmetic we let the computer form an error bound as it forms each numerical value. We are repaid for this extra computation by always having rigorous error bounds for all our computed answers.

2.4. RANGE ARITHMETIC

When interval arithmetic is combined with the variable precision system described in Section 2.2, we obtain the means of knowing when too few digits have been used in our computation. To the floating-point representation (2.2) we add a single decimal digit r and obtain the representation

$$(\text{sign}) \ 0.d_1 d_2 \cdots d_n \pm r \cdot 10^e \tag{2.7}$$

The digit r is the *range* digit, used to define the maximum error of the number, with the decimal significance of r identical to that of the last mantissa digit d_n. By adding r to d_n, or subtracting r from d_n, we form the endpoints of the mantissa *interval.* From the range digit we obtain the mantissa's error bound by taking account of the range digit's mantissa position and ignoring the exponent part. When this value is multiplied by 10^e, we obtain the number's error bound. Below these two values are given for some typical numbers.

Ranged number	Mantissa error bound	Number error bound
$+0.8888 \pm 9 \cdot 10^1$	$0.0009 = 9 \cdot 10^{-4}$	$9 \cdot 10^{-3}$
$-0.7244666 \pm 2 \cdot 10^{-2}$	$0.0000002 = 2 \cdot 10^{-7}$	$2 \cdot 10^{-9}$
$+0.2355555555 \pm 3 \cdot 10^4$	$0.0000000003 = 3 \cdot 10^{-10}$	$3 \cdot 10^{-6}$

When decimal constants are converted to this range arithmetic form, as before only as many mantissa digits appear as are needed to fully represent the constant, and the range digit is set to zero because the constant is without error. Thus the constant 2.5 would obtain the ranged form $+0.25 \pm 0 \cdot 10^1$. Ranged numbers like this one, with a zero range digit, are called *exact*.

When exact numbers now are added, subtracted, or multiplied, the result also is exact if the PRECISION bound is not overstepped. But if PRECISION is exceeded, then the mantissa gets truncated to the PRECISION length, and the number gets a range digit of 1, indicating a maximum error of one unit in the last mantissa digit. As before, the accuracy of the four rational operations is changed simply by changing PRECISION.

When an arithmetic operation is performed with one or both operands having a nonzero range digit, the result also gets a nonzero range digit, computed using the relations of the preceding section. Generally the result mantissa error bound is computed first, then rounded up to a one digit value to form the range digit. This fixes n, the number of mantissa digits. Then the result mantissa is formed to this length, and if this means some mantissa digits are discarded, then the result range digit is incremented by one.

As a computation proceeds, with the numbers now having nonzero range digits, the successive mantissas generally obtain increasing error bounds, and, accordingly, their length tends to decrease. In one respect this is helpful, in that suspect mantissa digits get discarded automatically. Shorter mantissas mean faster arithmetic operations, so usually the later part of a computation runs quicker than the earlier part. If the final results have too few mantissa digits to yield the required number of correct decimal places, then the whole computation is repeated at a higher PRECISION. The cases where this occurs are likely to be cases in which ordinary floating-point computation would yield inaccurate results. The amount to increase PRECISION is determined by examining each answer that is insufficiently accurate, to determine by how many digits it fails to meet its target form. If the greatest deficit in correct decimal places equals D, then

PRECISION is increased by D plus some small safety margin perhaps, and a repetition of the computation likely succeeds. Of course this description applies only to the simplest cases, where the computation is straightforward from exact initial constants to answers. The solution of a system of linear equations could serve here as an example. Usually the numerical solution of a mathematical problem is more complicated, and initial gross approximations to desired results are progressively refined. When we treat such cases we need some method of determining the error of the final approximants.

Range arithmetic is assumed now to be the arithmetic used in all the computations described in this text. Up to now we have displayed ranged numbers in the form used for machine computation, a form harder to read than the more familiar scientific floating-point form. In later sections we often convert displayed ranged numbers into a form resembling scientific floating point, moving the mantissa decimal point one digit to the right and decreasing the exponent by one. Thus instead of the computer form $+0.344445 \pm 1 \cdot 10^5$, for easier comprehension we may display the number as $3.44445 \pm 1 \cdot 10^4$, and instead of the computer form $+ 0.0 \pm 2 \cdot 10^{-3}$, we may display $0 \pm 2 \cdot 10^{-4}$. We also drop the exponent part if it equals one, that is, a number $-5.4311 \pm 2 \cdot 10^0$ is displayed simply as -5.4311 ± 2.

2.5. INTERVAL ARITHMETIC NOTATION

For a mathematical constant a, let us use $\underline{a}$ to denote its range arithmetic interval approximation obtained at the current PRECISION.

There are two ways to designate an interval: with the notation $m \pm \epsilon$ that we have been using and with the endpoint notation $[a, b]$. One notation is easily converted into the other: $m \pm \epsilon$ is also $[m - \epsilon, m + \epsilon]$, and $[a, b]$ is also $\frac{a + b}{2} \pm \frac{b - a}{2}$. Most often we use the $m \pm \epsilon$ notation, and we need names for the two parts. The number m is the *midpoint* of the interval, and because the width of the interval is 2ϵ, the number ϵ is the *halfwidth* of the interval. In this book, we often need to refer to the midpoint and halfwidth of various intervals that figure in a computation. To save space we use the abbreviations *mid* for midpoint and *wid* for halfwidth. Thus if b is a quantity for which the approximation $\underline{b}$ has been obtained, then b is somewhere in the interval

$$\text{mid } \underline{b} \pm \text{wid } \underline{b}$$

Often the mid or the wid of an interval must be examined for some reason. Suppose we have computed the quantity c in range arithmetic, obtaining the range arithmetic value

$$c = -0.12345 \pm 8 \cdot 10^{-2}$$

To generate mid c as another ranged number, all that needs to be done is to reset the range digit to zero. That is,

$$\text{mid } c = -0.12345 \pm 0 \cdot 10^{-2}$$

To generate wid c, the range of c is converted into a mantissa and assigned the correct exponent part. Thus continuing with this example, we would obtain

$$\text{wid } c = + 0.8 \pm 0 \cdot 10^{-6}$$

where the exponent is -6 after the wid $0.00008 \cdot 10^{-2}$ is converted to standard form.

2.6. COMPUTING STANDARD FUNCTIONS IN RANGE ARITHMETIC

After the arithmetic operations in range arithmetic have been coded with appropriate software, one can compose simple programs in this arithmetic. We often need, however, standard functions like $\ln x$, $\sin x$, or $\cos x$. For instance, we may need to find $\cos 3.12222 \pm 1$, where the cosine argument is in radians.

In range arithmetic, standard functions can be formed in much the same way they are formed with ordinary floating point, that is, by using power series expansions, in combination with various useful identities that apply to the particular function. Thus to form $\cos x$, we can use the three identities $\cos(-x) = \cos x$, $\cos x = \cos(x - 2\pi n)$, and $\cos x = -\cos(x - \pi)$ to replace the argument x by a value x' between 0 and $\pi/2$. Then we halve x' as many times as necessary to confine its new value x'' within a narrower interval, say $[0, 0.1]$. We record the number of halvings, intending to perform the scaling operation $y: = 1 - 2y^2$ on $y = \cos x''$ as many times as we halved, in accordance with the identity $\cos 2x = 1 - 2\cos^2 x$.

To compute $\cos x''$ we use the Maclaurin series expansion

$$\cos u = 1 - \frac{u^2}{2!} + \frac{u^4}{4!} - \frac{u^6}{6!} + \cdots + (-1)^n \frac{u^{2n}}{2n!} + \cdots$$

How many terms of the series should we use to form $\cos x''$? In the process of summing the series, we would be adding consecutive terms to a summation variable S. The terms rapidly decrease in magnitude, and when finally we come to a term T whose magnitude is less than the current value of wid S, then clearly this is a good point to break off the summation.

The exact number wid S bounds the computation error in forming the partial sum for $\cos x''$ but does not include the series truncation error that we make by ending the summation. Because the series is an alternating series, with successive terms decreasing in magnitude, the truncation error is bounded by the magnitude of the first series term not summed. We need to increase the range digit of S by an amount sufficient to account for the truncation error $|T|$.

This evaluation example shows that it is useful to introduce another operation to range arithmetic, the operation

$$S \oplus T$$

The result of $S \oplus T$ is the first operand S but with an increased wid, the increase being at least equal to the greatest possible $|T|$ value accomplished in the following manner. If $T = \text{mid } T \pm \text{wid } T$, then first the ranged number $T' = 0 \pm r \cdot 10^e$ is formed wid $r \cdot 10^e$ equal to or exceeding the value of $|\text{mid } T| + \text{wid } T$. Then an ordinary range arithmetic addition $S + T'$ is performed to get the result we want.

The operation $\oplus$ has many applications besides its use in standard function computation. When a mathematical constant is computed in range arithmetic, we get an interval with a wid that can be examined and tested. If we wish to print the constant to a certain number, say k decimal places, the size of the wid determines whether or not we can get k correct places from the mid. If we can not, the higher precision needed can be determined, and then the computation is repeated at this precision. The range arithmetic operations for doing this are described in Chapter 10.

A similar system is used when we want to solve some mathematical problem numerically. Here we may have one procedure for computing an approximation A to the answer and another procedure for computing an upper bound E on the error of our approximation. Both computations yield range arithmetic numbers with wids bounding their computational error. The operation $A \oplus E$ then gives a correctly ranged approximation to the answer, which can be tested to determine whether the required number of correct decimal places can be obtained from it. If not, we increase precision, recompute A and E, and obtain a better ranged approximation.

2.7. RATIONAL ARITHMETIC

A useful by-product of range arithmetic is that rational arithmetic becomes easy to obtain. For rational arithmetic we use a pair of exact numbers to represent each rational r, namely a numerator integer p and a denominator integer q to represent r in the form $\frac{p}{q}$. The denominator integer q can not be zero, but there is no restriction on the numerator p. The rational operations of addition, subtraction, multiplication, and division are executed according to the rules

$$\frac{p_1}{q_1} + \frac{p_2}{q_2} = \frac{p_1 q_2 + q_1 p_2}{q_1 q_2}$$

$$\frac{p_1}{q_1} - \frac{p_2}{q_2} = \frac{p_1 q_2 - q_1 p_2}{q_1 q_2}$$

$$\frac{p_1}{q_1} \times \frac{p_2}{q_2} = \frac{p_1 p_2}{q_1 q_2} \tag{2.8}$$

$$\frac{p_1}{q_1} \div \frac{p_2}{q_2} = \begin{cases} \dfrac{p_1 q_2}{q_1 p_2} & \text{if } p_2 \neq 0 \\ \text{Division error} & \text{if } p_2 = 0 \end{cases}$$

Before executing any of the four operations shown above, PRECISION is temporarily set to its highest value, N. The result p and q integers then are computed in range arithmetic, following the rules given above. Because no division operation is ever needed, the two integers obtained are always exact. If it should ever turn out that the range of one of these integers is nonzero, this signals that N, the implementation bound on the number of mantissa digits, is not large enough for the computation.

It is convenient to restrict denominator integers q to positive values only. It is then easier to decide whether we are dealing with a positive or negative rational, because we need examine only the sign of p, rather than the signs of both p and q. With this restriction, the equations for arithmetic operations given earlier must be reexamined. The equations for addition, subtraction, or multiplication need no changes because each of these operations yields a result with a positive denominator when both operands have positive denominators. This is not the case for division, however, and the revised operation is:

$$\frac{p_1}{q_1} \div \frac{p_2}{q_2} = \begin{cases} \dfrac{p_1 q_2}{q_1 p_2} & \text{if } p_2 > 0 \\ \dfrac{-p_1 q_2}{q_1 |p_2|} & \text{if } p_2 < 0 \\ \text{Division error} & \text{if } p_2 = 0 \end{cases}$$

It is convenient also to always eliminate common divisors from the pair of integers p, q, and obtain each rational in what is called *reduced* form. This way, we can conclude that a rational $\frac{p_1}{q_1}$ equals another rational $\frac{p_2}{q_2}$ after we check that p_1 equals p_2 and that q_1 equals q_2. If common divisors are not eliminated, we reach the same conclusion only after we find that $p_1 q_2$ equals $p_2 q_1$, and this requires two multiplications. Elimination of common divisors also keeps the integers p and q smaller, making arithmetic operations faster, and making it less likely that we exceed N, the bound on mantissa length. Accordingly, as each rational $\frac{p}{q}$ is obtained by an initial construction or through an arithmetic operation, the greatest common divisor D of $|p|$ and q is found, and if D is greater than 1, then the pair p, q is replaced by the pair p', q', with $p' = p/D$ and $q' = q/D$.

The method we use for finding D is the Euclidean algorithm, which finds the greatest common divisor of any two positive integers n_1 and n_2 by repeated divisions. The procedure is as follows: We divide the larger integer by the smaller to obtain an integer quotient and an integer remainder. Assume $n_1 \geq n_2$; if not, switch the integers to obtain this. Let d_1 be the quotient and n_3 the remainder when n_1 is divided by n_2, so that

$$n_1 = d_1 n_2 + n_3$$

with $0 \leq n_3 < n_2$. From the above equation it is clear that any common divisor of n_1 and n_2 is also a divisor of n_3, so the greatest common divisor of n_1 and n_2 is a common divisor of n_2 and n_3. On the other hand, any common divisor of n_2 and n_3 is also a divisor of n_1, so the greatest common divisor of n_2 and n_3 is a common divisor of n_1 and n_2. The two statements together imply that these two greatest common divisors are identical, so the problem of finding the greatest common divisor can be shifted from the pair n_1, n_2 to the pair n_2, n_3, and the entire process can be repeated on the new pair. We see that n_3, the smaller number of the new pair, is less than n_2, the smaller number of the original pair. Thus if this procedure is repeated enough times, it must lead eventually to a pair of integers with one member being zero. When this happens the desired greatest common divisor equals the nonzero member of this final pair. Thus for the integer pair 144 and 78, we arrive at their greatest common divisor in four steps:

Pair	Division Relation
144, 78	$144 = 1 \cdot 78 + 66$
78, 66	$78 = 1 \cdot 66 + 12$
66, 12	$66 = 5 \cdot 12 + 6$
12, 6	$12 = 2 \cdot 6 + 0$
6, 0	The greatest common divisor is 6

EXERCISES

1. Call `calc`, and when the program asks you to enter the number of decimal places you want, enter the single letter q and hit the <ENTER> key. All demo programs can be terminated whenever a keyboard entry is required by entering a lower case or upper case q (for quit). Call `calc` again, and this time enter a Control C character to stop execution. You should be able to terminate any demo program at any time by entering the Control C character, but this facility may not be provided by some C++ compilers.

2. Call `calc`, request three fixed-point decimal places, and then enter the constant `0.1111`. Note that a tilde appears after the three decimal result indicating it is correct to the last decimal place. The maximum error of any result appearing with a tilde is always plus or minus a 5 in the tilde position. Next enter the constant `0.11`, and notice that this time the tilde does not appear, indicating the constant is an exact rational. Enter the constants `sin(0)` and `sin(pi)`. The sin 0 entry leads to a display of an exact rational, and the sin π entry to a display with a tilde, even though both constants are zero. If a result known to be exact appears with an attached tilde, this indicates that when the result was computed it was obtained with a nonzero range digit.

3. Call `calc`, request 10 fixed-point decimal places, and calculate the quantities `exp(1)`, `log(exp(1))`, `exp(log(exp(1)))`, `log(exp (log(exp(1))))`, and so forth. Notice the slight delay to get the values of later entries, signifying the need for an increase in the precision of computation and a recalculation. (If you do not notice any delay because your computer is speedy, try 100 decimal places.)

4. Call `c_calc`, request 10 fixed-point decimal places, and calculate the complex quantities `exp(i*pi/2)`, `log(exp(i*pi/2))`, `exp(log (exp(i*pi/2)))`, `log(exp(log(exp(i*pi/2))))`, and so forth. Again notice the slight delay to get the values of later entries, signifying the need for an increase in the precision of computation and a recalculation.

5. Call `r_calc`, and calculate the quantities `2^10`, `2^100`, `2^1000`, `2^10000`. Here the delay for the later entries is due to the increased time of calculation and is not caused by recalculation, because precision for any `rational` operation is always the largest possible precision.

NOTES

N2.1. Five general texts on interval arithmetic [5, 8, 11, 12, 13] are listed below.

N2.2. The articles by Demmel and Krückeberg [6] and Krückeberg and Leisen [10] describe pioneering experiments in combining variable precision with interval arithmetic. Krückeberg [9] has given a general description of the advantages of this type of computation.

N2.3. Range arithmetic can be programmed as a binary system [1] or as a decimal system [2, 3, 4]. In the binary system the arithmetic operations execute faster, while in the decimal system the creation or display of a ranged number (input and output) execute faster and are easier to code.

N2.4. Another software package for variable precision interval arithmetic has been constructed by J. Ely [7].

REFERENCES

1. Aberth, O., A precise numerical analysis program, *Comm. ACM* **17** (1974), 509–513.

2. Aberth, O., Precise scientific computation with a microprocessor, *IEEE Transactions on Computers* **C-33** (1984), 685–690.

3. Aberth, O., *The conversion of a high order programming language from floating-point arithmetic to range arithmetic, in Computer Aided Proofs in Analysis, pp 1–6*, IMA Volumes in Mathematics and its Applications, Vol. 28, edited by K. R. Meyer and D. S. Schmidt, Springer-Verlag, New York, 1990.

4. Aberth, O. and Schaefer, M. J., Precise computation using range arithmetic, *ACM Trans. Mathematical Software* **18** (1992), 481–491.

5. Alefeld, G. and Herzberger, J., *Introduction to Interval Computation*, Translated by Jon Rokne, Academic Press, New York, 1983.

6. Demmel, J. W. and Krückeberg, F., An interval algorithm for solving systems of linear equations to prespecified accuracy, *Computing* **34** (1985), 117–129.

7. Ely, J. S., The VPI software package for variable precision interval arithmetic, *Interval Computations* **2** (1993), 135–153.

8. Hansen, E., *Global Optimization Using Interval Analysis*, Marcel Dekker, Inc., New York, 1992.

9. Krückeberg, F., *Arbitrary accuracy with variable precision arithmetic*, in *Interval Mathematics 1985*, edited by Karl Nickel, Lecture Notes in Computer Science 212, Springer-Verlag, Berlin, 1986.

10. Krückeberg, F. and Leisen, R., *Solving initial value problems of ordinary differential equations to arbitrary accuracy with variable precision arithmetic*, in Proceedings of the 11th IMACS World Congress on System Simulation and Scientific Computation, Vol. 1, Oslo, 1985.

11. Moore, R. E., *Interval Analysis*, Prentice-Hall, Englewood Cliffs, NJ, 1966.

12. Moore, R. E., *Methods and Applications of Interval Analysis, SIAM Studies in Applied Mathematics*, SIAM, Philadelphia, 1979.

13. Neumaier, A., *Interval Methods of Systems of Equations, Encyclopedia of Mathematics and its Applications*, Cambridge University Press, Cambridge, 1990.

SOLVABLE PROBLEMS AND NONSOLVABLE PROBLEMS

3.1. A KNOTTY PROBLEM

Imagine we are computing a certain quantity x to decide whether or not it equals $\frac{1}{2}$. If x equals $\frac{1}{2}$ we take branch EQUAL, otherwise we take branch UNEQUAL. The number x may be the value of a solution to a complicated differential equation problem at some particular point or of a number obtained through some other intricate computation. There is no difficulty in carrying out this program segment in ordinary floating point, but in this case the x approximation we generate has an unknown error, so we may end up taking the wrong branch.

Now reconsider the program fragment in range arithmetic. We list some sample cycles below. All exponent parts are $10^0 = 1$ and are omitted. Keep in mind that if x actually equals $\frac{1}{2}$, we can not expect to obtain the exact value 0.5 ± 0 when we compute $\underline{x}$, because this happens only for the simplest computations.

$\underline{x}$	Outcome
0.744 ± 2	branch UNEQUAL
0.50001 ± 3	recompute to higher precision
0.500000030 ± 4	branch UNEQUAL
0.50002 ± 4	recompute to higher precision
0.500000000 ± 2	recompute to higher precision
$0.500000000000000001 \pm 5$	recompute to higher precision
$\vdots$	$\vdots$

The problem with the last case is that we never are certain when to take branch EQUAL. Indeed, what should we do when we see a sequence of better and better approximations, defining smaller and smaller intervals, each containing the rational number $\frac{1}{2}$? We cannot continue to form better and better approximations indefinitely, because x *may* equal $\frac{1}{2}$, and then we would be in an endless loop. If we adjust our program fragment so that we take branch EQUAL when the x approximation's interval becomes "sufficiently" small, we really are taking branch EQUAL not just for $x = \frac{1}{2}$, but for x equal to any of the real numbers lying in the "sufficiently" small interval. Anyone examining our program fragment could demonstrate to us our error by arranging things so that x was not equal to $\frac{1}{2}$ but merely close enough so that it passed our criterion for taking branch EQUAL.

In our example we used the test constant $\frac{1}{2}$, which can be expressed as an exact ranged number, 0.5 ± 0. The difficulty takes a slightly different form if the test constant cannot be so expressed, for instance, if it is $\frac{1}{3}$ or $\sqrt{2}$. In this case both the test constant and the variable x require recomputation when their intervals overlap, as in the example below.

x	$\frac{1}{3}$	Outcome
0.3333 ± 2	0.3333 ± 1	recompute to higher precision
0.3333333333 ± 4	0.3333333333 ± 1	recompute to higher precision
0.333333333333333 ± 2	0.333333333333333 ± 1	recompute to higher precision
$\vdots$	$\vdots$	$\vdots$

If we were testing whether x equals 0, and we were unable to resolve the question, the successive x mantissas would not grow in length, but their exponent parts would be decreasing in magnitude, as in this example:

x	Outcome
$0.2 \pm 4 \cdot 10^{-5}$	recompute to higher precision
$0.0 \pm 2 \cdot 10^{-11}$	recompute to higher precision
$0.1 \pm 3 \cdot 10^{-18}$	recompute to higher precision
$\vdots$	$\vdots$

The point with all these cases is that the often seen phrase in an algorithm description "If $x = a$, then . . ." can not be handled correctly for all instances of a number x known precisely in that it can be computed as accurately as we please, but for which there is no additional information. The difficulty is not a peculiarity of range arithmetic but merely becomes evident with this mode of computation. With floating-point computation the difficulty was there too but is usually unrecognized or ignored; branch EQUAL is taken whenever the two floating-point approximations are identical, which of course does not imply that the two mathematical numbers being approximated are equal. This difficulty is *intrinsic* to the general problem of deciding whether or not two numbers are equal. In this text we draw attention to problems that have an intrinsic difficulty, the *nonsolvable* problems. The mathematical problems free of such difficulties are *solvable* problems. In the next section we give a brief introduction to the mathematical basis for this terminology. Our first classified problem is:

NONSOLVABLE PROBLEM 3.1: For any two numbers a and b, decide whether or not a equals b.

As soon as we have a nonsolvable problem P, any other problem whose solution would imply a method of solving P becomes nonsolvable too. For instance, as a consequence of Nonsolvable Problem 3.1, we have also

NONSOLVABLE PROBLEM 3.2: For any two numbers a and b, decide whether or not $a \circ b$, where $\circ$ may be one of the relations $>$, $\geq$, $<$, or $\leq$.

For instance, if it were possible to always determine whether or not $a > b$ were true, then from our *yes* or *no* results from testing $a > b$ and $b > a$, we would know whether or not a were equal to b, contradicting our first nonsolvable problem. Nonsolvable Problem 3.1 treats the decision problem for $a \circ b$ where $\circ$ is $=$ or $\neq$. Now we see similar results when $\circ$ is any of the four order relation symbols.

When precise computation is attempted, it is helpful to identify both solvable and nonsolvable problems. If a problem is identified as nonsolvable, one can try to change the problem is some acceptable way so that it becomes solvable. Thus with Problem 3.1, the difficulty can be avoided if we do not need to know whether or not the two numbers are equal but only whether or not they are *close* as in this sense:

SOLVABLE PROBLEM 3.3: For any two numbers a and b and any positive integer n, decide that a is within 10^{-n} of b, or decide that the two numbers are unequal.

The problem is solvable because we can compute a and b in range arithmetic to higher and higher precision until both approximations have a wid less than $\frac{1}{4} \cdot 10^{-n}$. Because the wid is the halfwidth of an approximation's interval, the sum of the lengths of the two intervals is less than 10^{-n}. So if the two approximation intervals overlap, the two numbers are within 10^{-n} of each other, otherwise the two numbers are unequal. Note that we do not try to decide *whether or not* a is within 10^{-n} of b, for that would be nonsolvable too, forcing us to decide, in some cases, whether or not $|a - b|$ was exactly 10^{-n}. The two choices listed in Problem 3.3 are not mutually exclusive. It is quite possible that at a certain precision we decide that a is within 10^{-n} of b, while at a higher precision we decide that a is unequal to b. That is, both decisions could be true, and then either choice is allowable.

Another way the difficulty of Problem 3.1 can be overcome, for inequality tests at least, is to have two distinct numbers to test against:

SOLVABLE PROBLEM 3.4: For any number a and any two unequal numbers b_1 and b_2, choose either b_1 or b_2 as a number unequal to a.

Because we know the numbers b_1 and b_2 are distinct, by computation to sufficiently high precision the ranged values of b_1 and b_2 do not overlap, and a certain rational width δ separates the two intervals. If we now compute a to a precision high enough that its wid is less than $\delta/2$, the a interval can overlap at most one of the b_1, b_2 intervals, and so at least one of the b_1, b_2 numbers can be determined to be unequal to a.

A third way the difficulty of Nonsolvable Problem 3.1 can be overcome is by putting some restriction on the type of numbers that are compared. Certainly if the two numbers are rational and are expressed in the form p/q with the integers p, q specified, there is no difficulty deciding any order relation, and generally, when all numbers are specified to be rationals, many nonsolvable problems become solvable. However, for order relations between a and b, it does not suffice merely to know that the two numbers are rational, if we have only range arithmetic approximations to them.

For real numbers a, b, there is exactly one of three possibilities: $a < b$, $a = b$, or $a > b$. We know from Nonsolvable Problem 3.1 that actually deciding which of the three is nonsolvable, and because this problem appears often, we list it below and a revised, solvable version.

NONSOLVABLE PROBLEM 3.5: For any two numbers a and b, select the single correct relation from these three: $a < b$, $a = b$, or $a > b$.

Using Solvable Problem 3.3, we have also

SOLVABLE PROBLEM 3.6: For any two numbers a and b and any specified positive integer n, select a correct relation from these three: $|a - b| < 10^{-n}$, $a < b$, or $a > b$.

Let us again denote a ranged approximation to a by $\underline{a}$ and a ranged approximation to b by $\underline{b}$. If the $\underline{a}$ interval overlaps the $\underline{b}$ interval we indicate this by writing $\underline{a} \doteq \underline{b}$. For instance, $2.22\pm2 \doteq 2.24\pm3$. Note that we count intervals as overlapping even if they just touch, so that $2.22\pm1 \doteq 2.24\pm1$. Similarly, we write $\underline{a} > \underline{b}$ if the $\underline{a}$ interval is to the right of the $\underline{b}$ interval, and we write $\underline{a} < \underline{b}$ if the $\underline{a}$ interval is to the left of the $\underline{b}$ interval. Thus $3.33\pm1 > 3.30\pm1$ and $3.33\pm1 < 3.36\pm1$. For any ranged approximations $\underline{a}$ and $\underline{b}$, we always have exactly one of three possibilities: $\underline{a} < \underline{b}$, $\underline{a} \doteq \underline{b}$, or $\underline{a} > \underline{b}$. From $\underline{a} < \underline{b}$, it follows that $a < b$, and from $\underline{a} > \underline{b}$, it follows that $a > b$, but from $\underline{a} \doteq \underline{b}$ it does not follow that $a = b$. Given $\underline{a} \doteq \underline{b}$, any of the relations $a < b$, $a = b$, or $a > b$ could be true. For ease in reading, from now on we usually do not use underlining when we give a dotted relationship. The reader should understand, however, that a dotted relationship always applies to ranged approximations of the mathematical quantities in the relationship obtained at some particular computation precision. So if $a \doteq b$ is true at a certain precision, then when precision is increased we may still find $a \doteq b$, but we could find $a > b$ or we could find $a \lessdot b$. It is convenient to define $a \neq b$ to mean either $a \lessdot b$ or $a > b$. Then it is always true that $a \doteq b$ or $a \neq b$, because it is always true that $a \lessdot b$, $a \doteq b$, or $a > b$. Similarly, we define $a \lesseqgtr b$ to mean $a \doteq b$ or $a \lessdot b$ and $a \gtreqless b$ to mean $a \doteq b$ or $a > b$. It may be useful here to list the conclusions that can be reached when various dotted relations are true.

Dotted relation	Implied mathematical relation
$a \lessdot b$	$a < b$
$a > b$	$a > b$
$a \neq b$	$a \neq b$
$a \doteq b$	None certain

Conversely, if we know that a certain mathematical order relation is true between two numbers, this information limits the dotted relations we could find.

Mathematical relation	Implied dotted relation
$a < b$	$a \lesseqgtr b$
$a > b$	$a \gtreqless b$
$a = b$	$a \doteq b$
$a \neq b$	None certain

In later chapters, in a description of a range arithmetic computation method, for any test to see whether a certain mathematical relation is true, we use dotted relations to describe the test. One should read "if $a \doteq b$" as "if $\underline{a}$ overlaps $\underline{b}$," and not as "if a equals b." As we hope to make clear in the next section, there is no way to test whether two quantities a and b are equal that works in all cases. With range arithmetic, we form a and b approximations at a specific precision and obtain one of three possible outcomes, $a \doteq b$, $a > b$, or $a \lessdot b$. Either outcome $a > b$ or $a \lessdot b$ implies $a \neq b$, but the outcome $a \doteq b$ implies only that a and b are *close*, specifically, that the distance separating a and b is not greater than twice the sum of the $\underline{a}$ and $\underline{b}$ wids. So if $a \doteq b$, how close a and b are to each other depends on the precision of computation.

3.2. THE IMPOSSIBILITY OF UNTYING THE KNOT

In the 1930s revolutionary methods were introduced into mathematics that have changed our understanding of the subject. The logician Gödel [3] proved that the foundations of mathematics will always be incomplete. Turing [5], in a brilliant paper defining the machines that have come to be known as Turing machines, proved that a certain long-standing problem, the Halting problem, was not solvable by any mechanical method. Both of these stunning findings were obtained by assuming that the result in question could be achieved and then by showing that this assumption leads to a contradiction.

It will be helpful to understand in general terms the methods Turing introduced to prove his result. The Halting problem is that of determining *by finite means* whether any given *finitely defined process* involving general integer computation terminates or not. This question is somewhat vague until one makes explicit the meaning of the terms *by finite means* and *finitely defined process*. To do this we can define a general-purpose computer and say that a process is finitely defined if the process is capable of being programmed on our computer. Our computer must be *ideal* in the sense that its memory must be unbounded, otherwise the size of this memory affects in a trivial way all our conclusions. The instruction set we assign to our computer is not particularly important, as long we give our computer certain basic capabilities. We are basing this statement on the fact that numerous alternate definitions of finitely defined process have been proposed and all such definitions have been found to be equivalent. One might say that one ideal computer can do, in its fashion, whatever any other ideal computer can do, in *its* fashion.

If the initial data used by a program P consists of the separate integers $a_1, a_2, \ldots, a_n$, then the result of its computation is denoted by $P(a_1, a_2, \ldots, a_n)$. Whatever the result is, it is deposited in memory in some convenient location, and the program terminates, its last computer instruction being a halt instruction. It is possible that $P(a_1, a_2, \ldots, a_n)$ is undefined because the program P with that particular input never terminates. The problem of determining by a general finite process whether any specified computation $P(a_1, a_2, \ldots, a_n)$ terminates is essentially the problem Turing addressed.

There is a general method of deciding by finite means whether any finitely defined process, with integer arguments $a_1, a_2, \ldots, a_n$, terminates, if there exists a program P^* that can examine programs P along with the initial data set $a_1, a_2, \ldots, a_n$ and determine whether or not the program P with this input terminates. That is, we suppose that $P^*(P, a_1, a_2, \ldots, a_n)$ always terminates after giving the result 1 or 0 according as $P(a_1, a_2, \ldots, a_n)$ does or does not terminate. A program for a modern computer, as it resides in memory storage before being executed, can be viewed as a binary integer with the most significant binary digit being the first storage byte's leftmost bit value and with the least significant binary digit being the last storage byte's rightmost bit value. Similarly, a program P for an ideal computer can be associated with a specific integer. So in the expression $P^*(P, a_1, a_2, \ldots, a_n)$, the entry P really is a certain integer from which all the steps of P can be determined.

When a finitely defined process and by finite means are made concrete in this way, by requiring expression as a program for an ideal computer, it becomes easy to see that deciding by finite means whether any given finite process terminates is impossible. Assuming the required program $P*$ is available, we obtain a contradiction. We can construct another program $P^{\sim}*$ such that $P^{\sim}*(a_1, a_2, \ldots, a_n)$ is computed as follows: the program $P*$, stored within the program $P^{\sim}*$, is extracted and used to compute $P*(P^{\sim}*, a_1, a_2, \ldots, a_n)$. If this result is 1, indicating termination, then $P^{\sim}*$ enters a loop; if this result is 0, indicating nontermination, then $P^{\sim}*$ terminates. Thus $P*$ is shown to fail for at least one program P, namely $P^{\sim}*$.

We have been vague here about all the details about our ideal computer: exactly what its list of instructions are, how a data set is represented in memory, in what way the computer's memory is infinite in extent, exactly how results of computation are communicated, and so forth. These details are of course important in any detailed proof, but here we are just trying to see in general terms the difficulties uncovered by Turing's line of reasoning. It is the assumption that it is possible, by a finite procedure, to decide certain things that nonterminating programs do that leads to contradictions. If, say, we assume that there is a general method of deciding whether or not a particular type of instruction ever gets used an infinite number of times by computations $P(a_1, a_2, \ldots, a_n)$, or that there is a general method of deciding whether or not such computations ever give a certain output an infinite number of times during computation, the same reasoning makes it clear that doing this with a finitely defined process is impossible.

It is important to realize that there is no difficulty in deciding by finite means anything about finite processes *that terminate*, because our decision can be made by stepping through the programs defining such processes to see what they do. For instance, suppose now that $P*$ is a program that determines something about computations $P(a_1, a_2, \ldots, a_n)$ *that are defined*, so that $P*(P, a_1, a_2, \ldots, a_n)$ terminates after giving a 1 or a 0 as its only output, depending, say, on whether or not P ever gives an output of 5. Again let us attempt to obtain a contradiction by defining $P^{\sim}*$ to be a program such that $P^{\sim}*(a_1, a_2, \ldots, a_n)$ is computed by extracting the stored program $P*$, determining the result $P*(P^{\sim}*, a_1, a_2, \ldots, a_n)$, and then behaving in a way opposite to what $P*$ predicts. This time, because $P*$ may be assumed to do its computation by stepping through the program $P^{\sim}*$, the program $P^{\sim}*$ enters an infinite loop, the computation alternating between checks of $P*$ by $P^{\sim}*$ and checks of $P^{\sim}*$ by $P*$. We see that $P^{\sim}*(a_1, a_2, \ldots, a_n)$ is undefined, and because $P*$ was required to decide questions only about defined computations, there is no contradiction.

Let us return to the question of deciding whether a program halts. There is no doubt that one can construct a program $P*$ that can correctly determine for a huge number of cases whether or not computations $P(a_1, a_2, \ldots, a_n)$ are defined. By analyzing at length typical programs of our ideal computer, one could no doubt make a long program $P*$ that would be very able. It could step through the supplied program P for an enormous number of steps to see whether it terminates, and if it did not find termination by then, it could try to detect the presence of a huge variety of program loops. But the Turing proof shows that such a program $P*$ must always be incomplete. There is an *intrinsic* difficulty to the problem of deciding by finite means whether finitely defined processes terminate, because the assumption that such a decision procedure exists immediately leads to a contradiction.

In the same paper in which Turing analyzed the Halting problem, he also considered which real numbers could be computed on his machines. Such numbers he called *computable*. In the early 1950s, a formal analysis treating computable numbers was founded by the Russian mathematician A. A. Markov and greatly extended by two other Russian mathematicians, I. D. Zaslavsky and G. S. Ceitin. In this analysis it is possible to prove the impossibility of solving computational problems in much the same way it is possible to prove the Halting problem is nonsolvable. This analysis, called *computable analysis*, is helpful in numerical analysis because it identifies computational problems that have intrinsic difficulties, that is, problems that are similar to the Halting Problem in that a general, finitely defined method of solving them is impossible. Such problems are called nonsolvable problems. If we assume a general solution method by finite means exists for a nonsolvable problem, immediately we are led to a contradiction.

To illustrate how this is done for computational problems, we sketch how Problem 3.1 is shown to be nonsolvable. In computable analysis each mathematical entity requires a finitely defined specification. Let us consider real numbers. One way of defining a real number is to give a convergent sequence of rational numbers approaching the number in the limit. Accordingly, for a real number a we require a function $\alpha(n)$ such that for any positive integer n, the rational number $\alpha(n)$, if converted to decimal form, is an n decimal-place a approximation to exactly n places, and such that its error does not exceed 10^{-n}. To be finitely defined, the function $\alpha(n)$ must be realizable on our ideal computer with a program P_α. That is, $\alpha(n) = P_\alpha(n)$. Here we are now allowing our programs P to give rational number outputs, but because a rational number $\frac{p}{q}$ can be defined as an ordered pair of integers p, q, there is no difficulty in allowing this. Simply think of each rational output as given by a pair of integers. Now suppose that there is a way of determining whether or not any finitely defined real number a equals zero. This is a special case of Problem 3.1, and if we show this simpler problem is nonsolvable, then Problem 3.1 must be too. So let $P*$ be a program that can decide whether any finitely defined real number equals zero. That is, $P*(P_\alpha)$ always terminates for an input program P_α defining a real number a and gives the single result 0 or 1 according as a is or is not zero. Because P_α is available to the decision program P^*, it can form arbitrarily many approximations $\alpha(n)$ to aid it in its decision. Now let $\alpha'(n) = P_{\alpha'}(n)$ behave as follows: for any n, using a stored copy of the program P^*, it follows the computation of $P^*(P_{\alpha'})$ through n steps to see if termination occurs. If termination does not occur by that number of steps, it gives an output of 0 as its value of $\alpha'(n)$. Thus its a approximation is an n decimal place 0.00 . . . 0, with error no greater than 10^{-n}. On the other hand, if it finds that $P^*(P_{\alpha'})$ terminates in exactly N steps, with of course $N \leq n$, then its output is 10^{-N} if P^* gives an output of 0 (meaning a is zero) and 0 if P^* gives an output of 1 (meaning a is not zero). Note that 10^{-N} equals 0.00 . . . 01 with the "1" in the Nth decimal place. Thus whatever P^* indicates, its result is in error. The problem of determining by finite means whether finitely defined real numbers are zero is shown to be nonsolvable.

Notice the insidious means by which the program P^* is thwarted. For its computation of $\alpha'(n)$, the program $P_{\alpha'}$ cannot simply go immediately to its stored program P^* and try to determine what P^* predicts for $P_{\alpha'}$, for then $P_{\alpha'}(n)$ no doubt would be undefined, and P^* is assumed to give outputs just for programs that are defined properly. So a waiting game is played, an output of 0 is routinely given for every n after a preliminary check is made of $P^*(P_{\alpha'})$ by stepping through n steps. This ensures that $P_{\alpha'}(n)$ is always defined. When for some perhaps extremely large n, the program $P_{\alpha'}$ "discovers" what P^* has determined about it, then for that n, and for all larger n, it gives an output value that does not conflict with its outputs for smaller n but a value that shows P^* is wrong. We can even imagine that the program P^* "discovers" the intent of $P^{\sim *}$. This does it no good, because it is supposed to be defined for all programs defining numbers, and so it must eventually give a decision for $P^{\sim *}$. After it does, its verdict, whatever it is, is used against it to prove it wrong!

This result shows there is an *intrinsic* difficulty in deciding by finite means whether finitely defined real numbers are zero. Even though it is logically clear that every finitely defined real number must be zero or not zero, we will never be able to decide by finite means whether *any* such number is zero. Any method we use can never cover all cases.

For reference, we list here the specific problem we showed to be nonsolvable.

NONSOLVABLE PROBLEM 3.7: For any number a, decide whether or not a equals zero.

There are many repercussions from this result. Any problem whose solution by finite means implies a method of treating Problem 3.7 becomes nonsolvable too, because the assumption of such a solution implies a result known to be self-contradictory. Thus Problem 3.1 is nonsolvable because a solution method here implies a solution method for Problem 3.7. Each problem added to the list of nonsolvable problems becomes itself a test for other problems, and these in turn become tests for other problems, and so on. Thus Problems 3.2 and 3.5 are nonsolvable because a solution method for either problem implies a solution method for Problem 3.1. Whenever we obtain nonsolvability results we are imagining the computation done with an ideal computer, with no limit to its memory, and no concern for the number of steps it takes to achieve a result. If a problem is nonsolvable, that does not mean that in *every* instance of the problem, we are going to encounter difficulty. For many nonsolvable problems, from a statistical perspective, in the great majority of cases actually encountered, there is no difficulty in obtaining a solution. But because the problem is rather like the Halting problem, there can be no *general*, finitely defined method of solving the problem. This means that if we attempt to create a general solution program, we are going to find cases, often fairly simple cases, in which the solution procedure fails, either by giving us an incorrect answer or by not terminating.

In a sense, when a problem is shown to be nonsolvable, this is a liberating result. We no longer need to wonder how to tackle the problem, because we see that any method we use can only be partially successful. It is more reasonable to consider how to avoid the problem altogether.

For almost every nonsolvable problem given in this text, a constructive way of solving the problem implies a contradiction to Nonsolvable Problem 3.7. Occasionally, to obtain a contradiction, we need a weaker form of Nonsolvable Problem 3.7, namely

NONSOLVABLE PROBLEM 3.8: For any number a, decide that $a \leq 0$ or decide that $a \geq 0$.

Note that we do not try to decide *whether or not* $a \geq 0$, which would be a form of Problem 3.2, but make only the weaker selection indicated. A program P^* that could solve this problem would give the result 0 or the result 1 according as $a \geq 0$ or $a \leq 0$, respectively, giving either result if $a = 0$. The thwarting program $a'(n)$ would for any n give the result 0.00 . . . 0 until it determined what P^* had indicated about itself, and then switch to either 0.00 . . . 1 or -0.00 . . . 1 to contradict P^*.

3.3. NONSOLVABLE PROBLEMS COMPARED WITH SOLVABLE PROBLEMS

When we call a mathematical problem nonsolvable, we are identifying as self-contradictory an imagined program on an ideal computer, the problem's general solution procedure. An attempt to solve a nonsolvable problem in general fashion is very much like an attempt to create a program to solve the Halting Problem. With more and more subroutines included to treat particular situations, more and more cases of the problem will be correctly handled. But there will always be cases for which the solution procedure fails.

Similar reasoning can be applied to other constructions besides solution procedures. For instance, for a proposed real-valued function $f(x)$ of a real variable x, it may turn out that realizing the function on an ideal computer is self-contradictory. In this case we call the function *not finitely realizable* instead of using the adjective *nonsolvable*. We use this term later in this chapter, in Section 3.5.

Now let us consider the other variety of problems, the solvable kind. For solvable problems the understanding is that all input quantities and functions satisfy the conditions set out in the problem. It is completely unspecified or "undefined" what occurs if an input not be as designated. We usually show a problem is solvable by describing a general solution method with range arithmetic, being initially unconcerned with the time taken for a solution. After the problem is shown to be solvable in this way, we can consider how to solve it efficiently.

If a problem is solvable, it is possible to devise a solution procedure to handle all instances of the problem on an ideal computer, but here the theory is indifferent to the number of steps taken. As a practical matter, we may fail to obtain an answer to a particular instance of a solvable problem when attempting its solution on present-day computers, because the required solution time may be too long. However, unlike a nonsolvable problem, simple instances of a solvable problem pose little difficulty. Usually parameters of the problem can be identified that determine the solution time. Thus with Problem 3.3, deciding whether two numbers a and b are within 10^{-n} of each other, we expect the solution time to depend on the size of n and the complexity of the programs used to compute a and b. One could imagine that the a and b programs are so complex that an answer to Problem 3.3 is not practically possible. Generally, with any solvable problem, there are instances of the problem that can not be solved in a reasonable amount of time with present-day computers. When the solution of a solvable problem is out of reach this way, we call the problem *computerbound*. The solution of a computerbound problem always depends on finding a more powerful machine, or if this is not feasible, on improving the solution algorithm. Computerbound problems of today may not be computerbound in the future, just as many problems previously computerbound are now easily solved with today's machines.

3.4. SOME SOLVABLE AND NONSOLVABLE DECIMAL-PLACE PROBLEMS

Previously, we identified the difficulty in deciding whether one number was equal to, less than, or greater than a second number. When ranged approximations to the two numbers yield overlapping intervals, and this keeps happening as we increase the precision, we are in a quandary as to which relation holds. This difficulty shows up sometimes when we attempt to obtain a correct k place fixed-decimal-point approximation to a number. For instance, suppose we are trying to get a correct five-place approximation to a certain number c, and we obtain the following sequence of ranged approximations to c:

$$0.1111150 \pm 2$$
$$0.1111150000 \pm 1$$
$$0.111115000000000 \pm 3$$
$$\vdots$$

We are unable to decide whether our correct five-place result should be $0.11111\tilde{\ }$ or $0.11112\tilde{\ }$, because the various c approximations all contain within their intervals the point 0.111115, which is midway between 0.11111 and 0.11112. If c is less than the midpoint value, the only correct five-place, fixed-decimal-point approximation is $0.11111\tilde{\ }$; if c is greater than the midpoint value, the only correct one is $0.11112\tilde{\ }$; and if c equals the midpoint value, either $0.11111\tilde{\ }$ or $0.11112\tilde{\ }$ can be used. Thus choosing which five-place approximation is correct hinges on deciding whether c is less than 0.111115, equal to 0.111115, or greater than 0.111115. None of the c approximations shown above allows us to make this determination. We have an instance of Nonsolvable Problem 3.5, and because it is clear the difficulty is independent of the number of decimal places desired, we have

NONSOLVABLE PROBLEM 3.9: For any real number c and positive integer k, obtain a correct k decimal-place, fixed-point approximation to c.

A way out of the impasse is easy to find: Whenever it becomes difficult to decide between two neighboring k place approximations, give one extra correct decimal place. This should not be difficult, because if the point $d_0.d_1d_2 \ldots d_k5$ is inside a succession of c approximation intervals, then $d_0.d_1d_2 \ldots d_k5$ is a correct $k + 1$ place approximation as soon as the wid of our c approximation is less than $\frac{1}{2} \cdot 10^{-(k+1)}$, the error bound for a correct $k + 1$ decimal-place approximation. For instance, with our example above, a six-place approximation, namely $0.111115\tilde{\ }$, can be obtained from the very first ranged value, where the difficulty first appeared. The error of this initial c approximation, two units in the seventh decimal position, is below the allowance for six correct places, namely 0.0000005. In general, we have

SOLVABLE PROBLEM 3.10: For any real number c and positive integer k, obtain a correct k decimal-place or a correct $k + 1$ decimal-place, fixed-point approximation to c.

In our C++ demonstration programs, when results are to be displayed to a certain number of correct decimal places in fixed-point format, an extra decimal place may occasionally be noticed for some answers. The extra place is always a "5" supplied automatically if the difficulty we have been describing shows up.

An interesting question is what error bound must we achieve, to approximate a quantity Q to either k or $k + 1$ correct decimal places in the manner allowed by Solvable Problem 3.10? We consider the case were k equals three. Suppose that we have obtained a rational approximation $R = d_0.d_1d_2d_3d_4d_5 \ldots d_n$ to Q, with ϵ being R's error bound. The bound ϵ is obtained by summing bounds for all errors made in forming R, including the computational error. We can form a three-place decimal approximation from R, but then we make an additional rounding error, which can be as large as $0.5 \cdot 10^{-3}$. Thus the rationals 0.111422, 0.1115, and 0.11065 all yield the three-place approximation $0.111\tilde{}$, with rounding errors of $0.422 \cdot 10^{-3}$, $0.5 \cdot 10^{-3}$, and $0.35 \cdot 10^{-3}$, respectively. Because we want our three-place value to be correct, ϵ plus the three-place rounding error for R cannot exceed $0.5 \cdot 10^{-3}$, which means the bound on ϵ is $0.5 \cdot 10^{-3}$ minus the rounding error. When the rounding error equals or is close to $0.5 \cdot 10^{-3}$, restricting the size of ϵ, we try to form a four-place approximation from R. Now the bound for ϵ becomes $0.5 \cdot 10^{-4}$ minus the four-place rounding error. If R is between 0.111475 and 0.111525, our four-place approximation is 0.1115 with a rounding error of no more than $0.25 \cdot 10^{-4}$, so the minimum value for ϵ is $0.25 \cdot 10^{-4}$. If R is between 0.1110 and 0.111475 or between 0.111525 and 0.112, a three-place approximation also gives a minimum value for ϵ of $0.25 \cdot 10^{-4}$. Thus for R varying between 0.111 and 0.112, the smallest allowable ϵ is $0.25 \cdot 10^{-4}$, occurring when R equals 0.111475 or 0.111525. In general, to be certain that we can form a correct k or $k + 1$ place approximation to a quantity Q, we must find a rational approximation to Q with the error bound

$$\epsilon_k = 0.25 \cdot 10^{-(k+1)} \tag{3.1}$$

We consider next the problem of forming correct k place scientific floating-point approximations. It is clear that the difficulty of forming exactly k decimal places can occur here too.

NONSOLVABLE PROBLEM 3.11: For any real number c and positive integer k, obtain a correct k decimal-place, scientific floating-point approximation to c.

Here there is even an additional difficulty that is sometimes encountered. Suppose we are attempting to find a correct four-place floating-point approximation to a number c, and the series of ranged approximations we obtain, at ever increasing precision, is as follows:

$$0.1 \pm 2 \cdot 10^{-5}$$
$$0.0 \pm 3 \cdot 10^{-12}$$
$$0.1 \pm 4 \cdot 10^{-17}$$
$$\vdots$$

The problem here is that we can not find even one correct decimal place if the zero point is in the c interval. If c is actually zero, we should display a "0" instead of any floating-point digits. Thus with scientific floating-point, when attempting to obtain a certain number of correct places, we may be bumping into the nonsolvable problem of determining whether or not a number equals zero. One way of revising Problem 3.11 to make it solvable is:

SOLVABLE PROBLEM 3.12: For any real number c and positive integer k, obtain a correct k decimal place or a correct $k + 1$ decimal-place, scientific floating-point approximation to c, or else indicate that $|c| \le 10^{-k}$.

Now we allow an *escape* from the decimal-place problem when we are uncertain whether c is zero. The magnitude bound 10^{-k} suggested here is arbitrary and can be replaced by some other convenient bound, for instance, 10^{-2k}, or 10^{-k^2}. By making the escape bound depend on k, we get smaller and smaller magnitude indications as more and more places are requested.

The escape magnitude bound used in our C++ system is one half the amount suggested in Problem 3.12. Suppose results are to be printed to five correct places in floating-point format. The escape is indicated by replacing the normal display, say 3.22333~ E12, by the display 0.~ E-n, where the integer n is always at least equal to k, the number of correct decimals requested. For instance, 0.~ E-5 or 0.~ E-9 may be displayed instead of a number in five-place floating-point format. A floating-point display normally starts with a nonzero integer, so displaying a leading zero makes the escape unmistakable. Because the error of 0.~ is at most one half of a unit in the 0 position, this makes the magnitude bound of 0.~ E-n to be $\frac{1}{2} \cdot 10^{-n}$.

In the remainder of this text, whenever we use the phrase *to k decimals* in the statement of a solvable problem, we mean *to k or k + 1 correct decimal-places in either fixed-decimal-point format or scientific floating-point format with a size escape allowed for floating point.*

Working toward k correct floating-point decimals is less convenient than working toward k correct fixed-place decimals, because the error bound we must achieve for a rational approximation R to a quantity Q is not known in advance of forming R. It is possible to use the error bound goal given in Equation 3.1 for $2k$ fixed-decimal-point places and then to form R subject to this restriction, because if we can display R to $2k$ or $2k + 1$ fixed-point places, then we also can display R to k or $k + 1$ floating-point places unless the R fixed-point display has the form

$$0.00 \cdot \ \cdot \ \cdot 0 \cdot \ \cdot \ \cdot$$

with k or more zeros following the decimal point. In this case we would use the floating-point escape. However, if it turns out that $|R|$ is large, then we have computed R more accurately than necessary. It may be more efficient to make a trial computation to determine R's floating-point exponent e. Once the exponent of R is known, the error bound

$$\epsilon_k = 0.25 \cdot 10^{-(k+1)} \cdot 10^e \tag{3.2}$$

may be used.

The difficulties in obtaining precisely k decimals for a number x do not occur if x is a rational number p/q and the integers p, q are known. Here when we form a k decimal-place approximation to x, our only error is the rounding error, which always can be kept to no more than $\frac{1}{2} \cdot 10^{-k}$.

SOLVABLE PROBLEM 3.13: For any rational number p/q and positive integer k, obtain a correct k place fixed-point approximation, and, if $p \neq 0$, a correct k place scientific floating-point approximation.

3.5. THE TROUBLE WITH DISCONTINUOUS FUNCTIONS

Suppose we have a certain real function $f(x)$ defined over an interval I. What should we require of $f(x)$ in order that it be *finitely defined*? Clearly from a finitely defined x argument lying in the domain I, we should be able to obtain a finitely defined $f(x)$ value. That is, if we can obtain arbitrarily accurate ranged approximations to x, than we should be able to obtain arbitrarily accurate ranged approximations to $f(x)$. Generally we need a computer routine whereby a ranged x approximation is used to compute a ranged $f(x)$ approximation. Further, the wid of the $f(x)$ approximation must decrease toward zero in some fashion as the wid of the x approximation decreases toward zero.

Now consider a function with a discontinuity at some argument. As a simple example we use the well-known sign function, usually designated by the abbreviation sgn. Its definition is

$$\text{sgn } x = \begin{cases} +1 & \text{if } x > 0 \\ 0 & \text{if } x = 0 \\ -1 & \text{if } x < 0 \end{cases}$$

This function has a discontinuity at $x = 0$. Ideally a routine realizing this function would return one of three exact values, $+1$, 0, or -1, according as it finds $x > 0$, $x = 0$, or $x < -1$, respectively. However, deciding which of the three x relations is true is Nonsolvable Problem 3.5. We can decide only whether $x > 0$,

$x \doteq 0$, or $x < 0$ at some finite precision. If $x > 0$ or $x < 0$, the value of sgn x is determined, but if $x \doteq 0$ and the x approximation interval contains positive and negative rationals, the ranged value for sgn x so far determined is 0 ± 1. Of course we can compute the x argument at higher and higher precisions, but if x is actually zero, more accurate x approximations all can be expected to have positive and negative rationals within their intervals. Only in the special case where we happened to obtain an exact zero for x, would we be able to supply an exact zero as the value for sgn x. So for most cases in which the argument x is zero, a better value for sgn x than 0 ± 1 never is determined. A finitely defined sgn function is not realizable. We see that the sgn x function we actually implement is

$$\text{sgn } x = \begin{cases} +1 & \text{if } x > 0 \\ \text{sometimes 0, but more} & \text{if } x = 0 \\ \quad \text{often undefined} & \\ -1 & \text{if } x < 0 \end{cases}$$

Let us call a mathematical idea a "nonrealizable concept" if it is not possible for an ideal computer to implement a constructive version of the idea. We have then

NONREALIZABLE CONCEPT 3.1: The sgn function defined for all numbers x.

It is clear that if such a function were realizable we would have a finite means for deciding whether any number was zero or not, contradicting Nonsolvable Problem 3.7. Just as with nonsolvable problems, in which we try to determine solvable versions, with a nonrealizable concept, it is illuminating to try to revise the idea to make it realizable. There is no difficulty realizing a sgn function defined for all numbers *except* zero, the point of discontinuity. For an arbitrary function argument x, our ideal computer forms range approximations to higher and higher precision endlessly, until at last it obtains $x \neq 0$, and then it returns a sgn value of $+1$ or -1 according as $x > 0$ or $x < 0$. Note that if a zero argument is supplied to our ideal computer, it enters an endless loop. If we wish, we could program the ideal computer to always check whether the x approximation is an exact zero and return the proper sgn value of 0 in that case. Still, this adjustment would not eliminate all the endless loops for possible zero arguments, and our function still would not be defined at $x = 0$.

REALIZABLE CONCEPT 3.2: The sgn function defined for all numbers x *except* zero.

The same difficulty is obtained with any function $f(x)$ with a discontinuity at some point x_0 within the function's domain. The discontinuity requires us to treat arguments x in such a way that we do one thing if $x = x_0$ and another thing if $x \neq x_0$. We can decide only whether $x \doteq x_0$ or $x \not\doteq x_0$, and this is insufficient.

Because of this difficulty, whenever in the statement of a solvable or nonsolvable problem some function is required as an input, *we assume the function is continuous throughout its domain*. Unless we do this, problems that require functions as part of their initial data become nonsolvable problems whenever function values must be determined to solve the problem.

In general, a mathematical function F with discontinuities is not finitely realizable. What is finitely realizable is a function equal to F at all points where F is continuous, and *undefined* at the points where F is discontinuous. *Throughout this text, any function mentioned in a theorem, in a solvable problem, or in a nonsolvable problem is to be presumed continuous in its domain.*

3.6. THE TROUBLE WITH FUNCTIONS AT DOMAIN ENDPOINTS

The demonstration program `calc` allows a user to enter, at the computer console, the number of correct decimal places wanted, and then, for each mathematical constant entered at the console, `calc` displays the value of the constant to this number of places. Suppose we use this program to find various square roots to five decimal places. If we enter `sqrt(4)` we get the correct answer, and if we enter `sqrt(-1)` we get an error message, both displays just as expected. But if we enter

$$\texttt{sqrt(sin(pi) - 10\^{}-90)} \tag{3.3}$$

we get the value `0.00000~`, and this is not correct! The number $\sin(\pi) - 10^{-90}$ equals -10^{-90}, so $\sqrt{\sin(\pi) - 10^{-90}}$ is not defined. The program `calc` should have given us an error message; instead it gave us a function value, a disappointing result. What is happening here?

We can view `calc` as attempting to solve for mathematical constants c the k or $k + 1$ decimal-place problem, Solvable Problem 3.10, so it may appear that `calc`'s failure is caused by some gross programming blunder. We can better understand the failure by following the evaluation of the entered constant. First $\sin \pi$ is computed to a certain precision, and a result, say $0 \pm 2 \cdot 10^{-18}$, is obtained. Then the ranged value $1 \pm 0 \cdot 10^{-90}$ is subtracted to obtain $0 \pm 3 \cdot 10^{-18}$, and this result is sent to the square-root routine, which returns a value, perhaps $0 \pm 2 \cdot 10^{-9}$, and this final result led to the incorrect five-decimal display. The trouble appears to be with the square-root routine.

Consider the function $\sqrt{x}$, which is defined for x in $[0, \infty)$. A range arithmetic square-root routine should return a value when it is called with an argument x that is not negative. If x is negative, say -0.2331 ± 1, then of course an error must be signaled. But what if the x argument is $0 \pm 4 \cdot 10^{-3}$, defining an interval stretching from -0.0004 to $+0.0004$? The true value of x lies somewhere within this interval, and x could be positive, zero, or negative. If x is negative, the only appropriate response is to signal an error. But if x is positive or zero, the negative part of the x interval can safely be ignored, and a square root can be computed that is valid for the rest of x's interval, $[0, 0.0004]$. The returned square-root value in this case could be $0 \pm 2 \cdot 10^{-2}$.

Consider an ideal computer realization of the square-root function such that the computer returns values only for arguments in $[0, \infty]$, rejecting every proffered argument that is outside this domain. Such a function is not realizable, because this requires determining whether or not $x \geq 0$, an instance of Nonsolvable Problem 3.2. It is possible to obtain an ideal computer realization of the square-root function, with the proper domain $[0, \infty]$, but then we must allow the computer to return values for some arguments x outside the domain.

The difficulty of the square-root function for x values near zero, appears with any function defined in some interval I, at an endpoint a of I. Thus with the functions $\sin^{-1} x$ or $\cos^{-1} x$ with domain $[-1, +1]$, if x equals 1.0000 ± 2 or -1.0000 ± 2, we are in a similar quandary as to whether to return a value or signal an error. There are two distinct ways of treating functions with a domain endpoint a, when the function, like the square-root function, is defined at the endpoint. Assume the endpoint is a left endpoint; the necessary changes for a right endpoint will be evident.

Endpoint In

With this approach, we evaluate incoming arguments x, and if we find $x \doteq a$ or $x > a$, we proceed with function evaluation. We signal an error if we find $x < a$. Using this approach, we have some improper function evaluations for arguments x less than a, but the function is defined correctly at the endpoint a and to the right of a.

Check for Endpoint

With this approach, we evaluate incoming arguments x, and only if we find $x > a$ do we proceed with function evaluation. We signal an error if we find $x < a$. If we find $x \doteq a$, we increase precision and evaluate x again. Using this approach, we have no improper function evaluations, but the function is undefined if $x = a$, because we enter an unending loop. Note especially that if we had followed this approach, the `calc` error with the constant (3.3) would not have occurred. Precision would have been increased again and again until it was high enough so that when $\sin \pi$ was computed, a result such as $0 \pm 2 \cdot 10^{-93}$ would be obtained. Now when the constant

$1 \pm 0 \cdot 10^{-90}$ is subtracted, a negative result $-1.000 \pm 2 \cdot 10^{-90}$ is obtained, and this time we get an error message when a square root is attempted.

Because we want all standard functions to be defined at their domain endpoints, our only recourse is to take the Endpoint In option. But this means that sometimes a function value is returned when an error indication should have been given.

The solvability analysis we give for mathematical problems is a means of detecting self-contradictory requirements. Specific assumptions are made, to be then analyzed for logical consistency. The `calc` program may be considered as attempting to solve Problem 3.10, which has no logical inconsistencies. Consider an ideal computer's solution of this problem, and suppose the constant c supplied to the ideal computer in the form of a program P_α for computing c approximations is faulty, and that the program P_α does not really do what it is supposed to do. What happens next is "undefined," because no specific assumptions were made about this case. Similarly with `calc` the failure to detect every improperly specified mathematical constant cannot be construed a failure in treating Problem 3.10. The detection of improperly specified mathematical constants is a separate issue, and it is true `calc` has failed in that regard. However, notice that it is a nonsolvable problem to determine by finite means whether a program P_α for an ideal computer really does define a number. In the nonsolvability proof sketched in Section 3.2 for Problem 3.7, simply change the problem from determining whether a number is zero to the problem of determining whether a proposed program really does define a number. The only change needed in the proof is to make P̄* enter a loop if it finds $P*$ claims c is a number and to continue giving 0 outputs if it finds $P*$ claims c is not a number. So an ideal computer cannot in all cases detect improper number arguments. Similarly it would be futile to try to make `calc` detect all erroneous mathematical constants.

Whenever a solvable problem has one or more real numbers as input data, there arises the possibility that one of these numbers is improperly specified and that the error is not caught. Perhaps the number is specified in terms of some standard function, and an out-of-domain argument extremely close to a domain endpoint is used. Similarly, if a solvable problem has one or more functions as input data, because a function may itself be a constant or may employ various constants, there arises the possibility that the function is incorrectly specified and that function values are obtained when an error indication should have been given. Our demonstration programs exhibit this behavior. Incorrectly specified constants and functions are usually identified as improper, but it is always possible to get by defenses. In summary, if a program purporting to treat a solvable problem fails to detect an error in problem specification, this cannot be considered a failure to meet the requirements of the solvable problem.

We return now to the `calc` program, which we supposed set to give five correct decimal places. If we enter at the console the constants

$$\texttt{ln(sin(pi) + 10\^-40)} \quad \text{or} \quad \texttt{1/(sin(pi) + 10\^-40)} \quad (3.4)$$

we get an error message! But $\sin \pi + 10^{-40}$ is positive, so we should have obtained a value for both entered constants. This seems to be poor performance again but of a different variety.

The difficulty this time is that the precision of computation is set too low. The interval computed for $\sin \pi + 10^{-40}$ includes the zero point, so both the logarithm routine and the division routine must indicate an error. If we exit `calc` and call the program anew but with 50 decimals specified, the precision of computation is set higher by `calc`, and now we obtain a correct result for either constant.

It is helpful to imagine the approach an ideal computer could use for the solvable problem of constant evaluation to a specified number of decimal places. If the computed logarithm argument is negative or an exact zero, then an error must be indicated. But if the argument interval includes both positive and negative numbers, then an error is uncertain, so precision must be increased and the computation repeated. With this approach the ideal computer could evaluate correctly all properly posed constants. If the `calc` program had followed a similar procedure, the constants (3.4) would have been correctly evaluated, but then the entered constants

$$\texttt{ln(sin(pi))} \quad \text{or} \quad \texttt{1/sin(pi)} \quad (3.5)$$

would make the program enter a loop, giving no messages of any kind. This is because $\sin \pi$ always evaluates to an interval containing the zero point, and this leads to endless computation cycles with steadily increasing precision. This illustrates the drawback of automatically increasing precision whenever an ambiguous error is encountered. An ambiguous error is one that prevents further computation but that may be resolved at a higher precision. Following the ideal computer approach has the by-product of replacing the identification of some errors by a program loop.

We prefer that our programs avoid infinite loops if at all possible, so precision is not automatically increased for an ambiguous error. Instead the appropriate "ambiguous error" message is given and computation halts. Thus for the left-hand entries of lines 3.4 or 3.5, the error message is that the logarithm routine received an argument $\doteq 0$; for the right-hand entries the error message is that the division routine received a divisor $\doteq 0$. This leaves the interpretation of the error and the action to be taken (that is, whether the problem should be reattempted at a higher precision) up to the program user.

EXERCISES

1. Call `calc`, request four fixed-point decimal places, and enter the quantities $1+5e-4$, $1+5e-5$, and $1+5e-6$, equal to 1.0005, 1.00005, and 1.000005, respectively. Notice that four decimal places are displayed for the first and third entries, with five decimals showing for the middle entry. Compare these displays with those obtained for the quantities $1+(5e-4)*\sin(pi/2)$, $1+(5e-5)*\sin(pi/2)$, and $1+(5e-6)*\sin(pi/2)$, which match the previous quantities. The first two displays no longer designate exact quantities, because this time zero range digits are not obtained.

2. Call `calc`, request five floating-point decimal places, and enter the quantities `sin(0)` and `sin(pi)`. Though both quantities designate a zero, only for the first entry is an exact 0 displayed, because the sine routine returns an exact zero only when the argument supplied is an exact zero. Note the floating-point escape that appears for `sin(pi)`. Rerun `calc` with 20 floating-point decimal places requested to see the changed escape display for `sin(pi)`.

3. Call `calc`, request five fixed-point decimal places, and enter the quantity mentioned at the beginning of Section 3.6, $\text{sqrt}(\sin(pi)-10^{-90})$. Note the incorrect result for this entry. Also note that the entry $\text{sqrt}(0-10^{-90})$ leads to an error message, and an error message also is obtained for the original entry if the number of decimal places requested is 50.

4. Call `calc`, request five fixed-point decimal places, and enter the quantity $\text{asin}(1+10^{-80})$, which is not defined because the arc sine argument is outside of $[-1, +1]$. An incorrect five decimal value is returned; the arc sine routine not detecting the out-of-bounds argument. The error is caught for an entry $\text{asin}(1+10^{-10})$. Also the error in the original entry is caught if the number of decimal places requested is 50.

5. Call `calc`, request five fixed-point decimal places, and enter the quantities $\ln(\sin(pi)+10^{-40})$ and $1/(\sin(pi)+10^{-40})$ given on line 3.4 of Section 3.6. Note the error message when a five decimal value is possible. These entries are evaluated correctly if a 0 is substituted for `sin(pi)`. The original entries are evaluated correctly if 50 decimal places are requested instead of five.

NOTES

N3.1. Turing's fundamental paper [5] and Gödel's incompleteness theorem [3] are discussed in Davis's book [2].

N3.2. Two books on computable analysis, [1] and [4], are listed below.

REFERENCES

1. Aberth, O., *Computable Analysis*, McGraw-Hill, New York, 1980.
2. Davis, M., *Computability and Unsolvability*, McGraw-Hill, New York, 1958.
3. Gödel, K., Über formal unentscheidbare Sätze der Principia Mathematica und vervandter Systeme I, *Monatshefte für Mathematik und Physik* **38** (1931), 173–198.
4. Kushner, B. A., *Lectures on Constructive Mathematical Analysis*, Translations of Mathematical Monographs, Vol. 60, American Mathematical Society, Providence, 1980.
5. Turing, A. M., On computable numbers, with an application to the Entscheidungsproblem, *Proceedings of the London Mathematical Society, series 2* **42** (1937), 230–265.

IV

COMPUTING DERIVATIVES AND INTEGRALS

4.1. COMPUTING THE DERIVATIVE OF A FUNCTION

Consider the following function with the parameter a.

$$f_a(x) = \begin{cases} a \sin \dfrac{x}{a} & \text{if } a \neq 0 \\ 0 & \text{if } a = 0 \end{cases} \tag{4.1}$$

The function $f_a(x)$ can be easily computed in range arithmetic as accurately as we please by using this procedure:

$$f_a(x) = \begin{cases} a \sin \dfrac{x}{a} & \text{if } a \neq 0 \\ 0 \oplus a & \text{if } a \doteq 0 \end{cases}$$

When $a \doteq 0$, it is safe to take f_a as $0 \oplus a$, because the interval obtained contains the zero point and approximates both 0 and $a \sin \frac{x}{a}$. If the accuracy of the f_a approximation is too low, then precision is increased and a new f_a value is obtained. As precision increases and we continue to find $a \doteq 0$, the width of the interval $0 \oplus a$ decreases, so we can obtain as accurate a value for $f_a(x)$ as we want without having to resolve the question whether or not a equals zero, a nonsolvable problem.

On the other hand, if we wanted to find the derivative of f_a at some point, there might be difficulty. If a is zero, $f'_a(x) = 0$, while if a is unequal to zero, $f'_a(x) = \cos \frac{x}{a}$. We see then that if $a \neq 0$, then $f'_a(0) = 1$, while if $a = 0$, then $f'_a(0) = 0$. There can be no general, finitely defined, procedure for finding the derivative of this function at the point $x = 0$, for if we had such a procedure, from $f'_a(0)$ we would be able to determine whether or not the real number a was zero, and this is a nonsolvable problem. Accordingly, we have

NONSOLVABLE PROBLEM 4.1: Given a function $f(x)$ known to have a derivative at a point x_0, find to k decimal places $f'(x_0)$.

There can be no general, finitely defined method of accurately determining derivatives of a function known to have derivatives from function values only. The usual procedure for finding an approximation to $f'(a)$ by computing $\frac{f(a + h) - f(a)}{h}$ requires us to know the proper h spacing to use, and as the preceding example shows, there is no general way of determining that just from the $f(x)$ function values.

Generally to accurately calculate the derivative of a function at some domain point, we need additional information beyond accurate function values.

4.2. THE ELEMENTARY FUNCTIONS

In a calculus course the great majority of functions used in the examples and problems can be expressed in terms of standard functions, such as $\sin x$ or e^x (exp x). Such functions we call *elementary functions,* and for these functions it is easy to calculate derivatives and to find various power series. As standard functions we include the two functions max(x, y) and min(x, y), which equal the maximum value and minimum value, respectively, of the two arguments x, y. With range arithmetic the only difficulty in computing max(x, y) occurs when $x \doteq y$. In this case for max(x, y) we choose the x or the y approximation with the greater right interval endpoint; in case of a tie, either approximation can be chosen. Similar considerations apply to min(x, y).

DEFINITION 4.1: A function of a finite number of variables $x, y, z, \ldots,$ is an *elementary* function if it can be expressed in terms of variables and constants using a finite number of the binary operations of addition, subtraction, multiplication, division, and exponentiation, the operation of single argument function evaluation with sin, cos, tan, $\sin^{-1}$, $\cos^{-1}$, $\tan^{-1}$, exp, or ln, and the operation of the double argument function evaluation with max or min.

A prefix minus sign, as in $-x$, can be viewed as a special case of a subtraction operation with an implied zero operand, that is, $-x = 0 - x$. The exponentiation operation with operands A, B produces A^B. The function abs(x) = $|x|$ is an elementary function, because it can be expressed either as max(x, $-x$) or in terms of exponentiation as $\sqrt{x^2} = (x^2)^{1/2}$. Sgn x is elementary also, as long we understand it to be undefined at $x = 0$, for we have sgn $x = x/|x|$.

In our list of standard functions, max(x, y) and min(x, y) are included for the sake of convenience. These two functions would be elementary even if dropped from our list. For we have max(x, y) $= \frac{1}{2}(x + y) + \frac{1}{2}|x - y|$ and min(x, y) $= \frac{1}{2}(x + y) - \frac{1}{2}|x - y|$, and, as mentioned, $|x - y|$ can be expressed in terms of exponentiation.

Consider a typical elementary function

$$f(x) = \frac{e^{-x} \cos 3.2x + 0.56x^4}{1 + \tan^2 3.15x} \tag{4.2}$$

To evaluate the function on a computer, we need a list of operations to be done. In the particular case above, the list is given in Table 4.1. In this table a symbol O_i represents the result of operation i.

In an evaluation list, the operands for an operation are always constants or variables, or results of previous operations. Constant and variable operands are "primitive" operands, the other operands are "derived". Simply by setting the variables x, y, z, . . . equal to specific values x_0, y_0, z_0, . . . , and then executing the operations of the evaluation list, we obtain the function value $f(x_0, y_0, z_0, . . .)$.

When we define an elementary function in some programming language, it is the language's compiler program that constructs the sequence of operations to use in evaluating the function. We need only adhere to the syntax of the language, and the compiler does the rest. Thus the function (4.2) might be defined in a hypothetical language by writing it as

TABLE 4.1　Evaluation List for the Function $f(x)$ of Eq. (4.2)

Operation	Type	Operand 1	Operand 2
1	Unary $-$	0	x
2	Exp evaluation	O_1	
3	$\times$	3.2	x
4	Cos evaluation	O_3	
5	$\times$	O_2	O_4
6	Exponentiation	x	4
7	$\times$	0.56	O_6
8	$+$	O_5	O_7
9	$\times$	3.15	x
10	Tan evaluation	O_9	
11	Exponentiation	O_{10}	2
12	$+$	1	O_{11}
13	$\div$	O_8	O_{12}

```
(exp(-x)*cos(3.2*x)+0.56*x^4)/(1+tan(3.15*x)^2)
```

We will see that with elementary functions the list of evaluation operations is useful in other applications besides just finding the value of the function. Because of this general use, whenever we use the term *elementary function,* we assume that a corresponding list of evaluation operations is available or can be obtained.

4.3. COMPUTING DERIVATIVES AND POWER SERIES OF AN ELEMENTARY FUNCTION $f(x)$

Consider an elementary function $f(x)$ with the power series (or Taylor series)

$$f(x) = a_0 + a_1(x - x_0) + a_2(x - x_0)^2 + \cdots + a_k(x - x_0)^k + \cdots \qquad (4.3)$$

The point x_0 is the *series expansion point,* and a coefficient a_k equals $\dfrac{f^{(k)}(x_0)}{k!}$.

We can use the evaluation list of $f(x)$ to generate the series 4.3 just by reinterpreting all the evaluation operations. By obtaining the power series 4.3, we also find derivatives at the series expansion point x_0, since $f^{(k)}(x_0) = k!a_k$. Suppose we want to form the series up to the term in $(x - x_0)^Q$. Each operand in an evaluation step may be considered to represent some intermediate function $g(x)$ with a power series

$$g(x) = g_0 + g_1(x - x_0) + g_2(x - x_0)^2 + \cdots + g_k(x - x_0)^k + \cdots$$

So now an operand supplies, instead of a number as previously, the set of power series coefficients $g_0, g_1, g_2, \ldots, g_Q$. Because $g_0 = g(x_0)$, we see that our previous use of the evaluation list to obtain function values can be viewed as generating single-term power series.

We need to make new interpretations for the primitive operands of $f(x)$'s evaluation list, that is, x and constants c. Clearly, for constants c we have

$$c = c + 0(x - x_0) + 0(x - x_0)^2 + \cdots$$

so a constant supplies the coefficient array $c, 0, 0, \ldots, 0$. For the variable x we have

$$x = x_0 + 1(x - x_0) + 0(x - x_0)^2 + \cdots$$

so the variable x supplies the array $x_0, 1, 0, 0, \ldots, 0$.

The five binary operations of an evaluation list need appropriate reinterpretations as series operations. Let g_k and h_k denote the coefficients of certain intermediate functions $g(x)$ and $h(x)$. For addition, subtraction, and multiplication, the series operation is clear:

$$(g + h)_k = g_k + h_k \tag{4.4}$$

$$(g - h)_k = g_k - h_k \tag{4.5}$$

$$(g \cdot h)_k = \sum_{i=0}^{k} g_i h_{k-i} = \sum_{i=0}^{k} g_{k-i} h_i \tag{4.6}$$

For division we have

$$\sum_{i=0}^{\infty} (g/h)_k (x - x_0)^k = \sum_{k=0}^{\infty} g_k (x - x_0)^k \Big/ \sum_{k=0}^{\infty} h_k (x - x_0)^k$$

and so

$$\sum_{k=0}^{\infty} g_k (x - x_0)^k = \sum_{k=0}^{\infty} (g/h)_k (x - x_0)^k \cdot \sum_{k=0}^{\infty} h_k (x - x_0)^k$$

Employing the multiplication relation (4.6), we get

$$g_k = \sum_{i=0}^{k} (g/h)_i h_{k-i} = \sum_{i=0}^{k-1} (g/h)_i h_{k-i} + (g/h)_k h_0$$

which leads to a recursive relation for the coefficients $(g/h)_k$:

$$(g/h)_k = \begin{cases} g_0/h_0 & \text{if } k = 0 \\ \dfrac{1}{h_0} \left(g_k - \displaystyle\sum_{i=0}^{k-1} (g/h)_i h_{k-i} \right) & \text{if } k > 0 \end{cases} \tag{4.7}$$

For the last operation of exponentiation, g^h, there are two procedures according as the operand h is or is not a constant. If h is not a constant, then we have

$$(g^h)_k = (e^{\ln g \cdot h})_k \tag{4.8}$$

and our exponentiation operation is converted to one multiplication and two function evaluations. The series operations for function evaluations are given further below. If h is the constant a, then we have

$$\frac{d}{dx} g(x)^a = a g(x)^{a-1} g'(x) \tag{4.9}$$

or

$$g(x) \frac{d}{dx} g(x)^a = a g(x)^a g'(x)$$

Multiplying by $(x - x_0)$ and converting to power series, we get

$$\sum_{k=0}^{\infty} g_k(x - x_0)^k \cdot \sum_{k=1}^{\infty} k(g^a)_k(x - x_0)^k = a \sum_{k=0}^{\infty} (g^a)_k(x - x_0)^k \cdot \sum_{k=1}^{\infty} kg_k(x - x_0)^k$$

This leads via relation (4.6) to the following equation for the coefficient of $(x - x_0)^k$:

$$\sum_{i=0}^{k} g_i(k - i)(g^a)_{k-i} = a \sum_{i=1}^{k} (g^a)_{k-i} i g_i$$

or

$$g_0 \, k(g^a)_k = \sum_{i=1}^{k} (g^a)_{k-i}(ai - (k - i))g_i$$

We obtain from this the recurrence relations

$$(g^a)_k = \begin{cases} g_0^a & \text{if } k = 0 \\ \dfrac{1}{g_0} \sum\limits_{i=1}^{k} \left(\dfrac{(a + 1)i}{k} - 1 \right) (g^a)_{k-i} g_i & \text{if } k > 0 \end{cases} \qquad (4.10)$$

Note that if $g_0 \doteq 0$, then because of the division by g_0, we cannot form the terms $(g^a)_k$ for $k > 0$. If a is a positive integer, the power series expansion exists, and even if a is not an integer, but positive, the terms of the series are defined for $k < a$. An alternate method of obtaining the series terms is needed. If a is a small positive integer we can use the relation

$$g^a = \underbrace{g \cdot g \cdot \cdots \cdot g}_{a \text{ factors}} \qquad (4.11)$$

and obtain the series by repeated multiplication. For the other cases we can obtain a method of specifying the terms from Equation 4.9. Multiplying this equation by $(x - x_0)$ and equating the coefficients of $(x - x_0)^k$, we obtain

$$k(g^a)_k = a \sum_{i=1}^{k} (g^{a-1})_{k-i} i g_i$$

so that

$$(g^a)_k = \begin{cases} g_0^{a-1} g_0 & \text{if } k = 0 \\ \dfrac{a}{k} \sum\limits_{i=1}^{k} (g^{a-1})_{k-i} i g_i & \text{if } k > 0 \end{cases} \qquad (4.12)$$

With this formula we can construct the terms $(g^a)_0, (g^a)_1, \ldots, (g^a)_Q$ from the terms $(g^{a-1})_0$, $(g^{a-1})_1$, . . . , $(g^{a-1})_{Q-1}$. This means we can start with just the single term $(g^{a-Q})_0 = g_0^{a-Q}$ of the function (g^{a-Q}) and then generate in succession the terms for the functions (g^{a-Q+1}), (g^{a-Q+2}), . . . , each time finding one more term then previously, until we arrive at the full list of terms for (g^a).

Next we reinterpret the evaluation operation for each standard function, the natural exponential function first. We have

$$e^{g(x)} = \sum_{k=0}^{\infty} (e^g)_k (x - x_0)^k$$

Taking derivatives, we get

$$\frac{d}{dx}(e^{g(x)}) = e^{g(x)}g'(x)$$

and if we multiply this equation by $(x - x_0)$ and express it in terms of power series, we get

$$\sum_{k=1}^{\infty} k(e^g)_k(x - x_0)^k = \sum_{k=0}^{\infty} (e^g)_k(x - x_0)^k \cdot \sum_{k=0}^{\infty} kg_k(x - x_0)^k$$

Employing equation (4.6) again, we see that for $k > 0$ we have

$$k(e^g)_k = \sum_{i=1}^{k} ig_i(e^g)_{k-i}$$

and again we obtain a recursive relation for the coefficients $(e^g)_k$:

$$(e^g)_k = \begin{cases} e^{g_0} & \text{if } k = 0 \\ \dfrac{1}{k}\sum_{i=1}^{k} ig_i(e^g)_{k-i} & \text{if } k > 0 \end{cases} \qquad (4.13)$$

For the natural logarithm we have

$$\frac{d}{dx}\ln g(x) = \frac{g'(x)}{g(x)}$$

After multiplying by $(x - x_0)$ and converting to power series, we get

$$\sum_{k=1}^{\infty} k(\ln g)_k(x - x_0)^k = \sum_{k=1}^{\infty} kg_k(x - x_0)^k \bigg/ \sum_{k=0}^{\infty} g_k(x - x_0)^k$$

On the left the constant coefficient is zero, so the relation 4.7 yields for the coefficient of $(x - x_0)^k$ the equation

$$k(\ln g)_k = \frac{1}{g_0}\left(kg_k - \sum_{i=1}^{k-1} i(\ln g)_i g_{k-i}\right)$$

Note here that when $k = 1$ the equation is $1(\ln g)_1 = g_1/g_0$ in accordance with the convention that the summation Σ is void when the stopping index is less than the starting index. After accounting separately for $(\ln g)_0$, we get

$$(\ln g)_k = \begin{cases} \ln g_0 & \text{if } k = 0 \\ \dfrac{1}{g_0}\left(g_k - \dfrac{1}{k}\displaystyle\sum_{i=1}^{k-1} i(\ln g)_i g_{k-i}\right) & \text{if } k > 0 \end{cases} \qquad (4.14)$$

From the relations

$$\frac{d}{dx}\sin g(x) = \cos g(x) \cdot g'(x)$$

$$\frac{d}{dx}\cos g(x) = -\sin g(x) \cdot g'(x)$$

we obtain in similar fashion the relations

$$(\sin g)_k = \begin{cases} \sin g_0 & \text{if } k = 0 \\ \dfrac{1}{k}\displaystyle\sum_{i=1}^{k} ig_i(\cos g)_{k-i} & \text{if } k > 0 \end{cases}$$

$$(\cos g)_k = \begin{cases} \cos g_0 & \text{if } k = 0 \\ -\dfrac{1}{k}\displaystyle\sum_{i=1}^{k} ig_i(\sin g)_{k-i} & \text{if } k > 0 \end{cases} \qquad (4.15)$$

Equations 4.15 make it clear that it is necessary to generate both sets of coefficients, $(\sin g)_k$ and $(\cos g)_k$, even when only one set is required.

The hyperbolic functions have similar power series relations

$$(\sinh g)_k = \begin{cases} \sinh g_0 & \text{if } k = 0 \\ \dfrac{1}{k}\displaystyle\sum_{i=1}^{k} ig_i(\cosh g)_{k-i} & \text{if } k > 0 \end{cases}$$

$$(\cosh g)_k = \begin{cases} \cosh g_0 & \text{if } k = 0 \\ \dfrac{1}{k}\displaystyle\sum_{i=1}^{k} ig_i(\sinh g)_{k-i} & \text{if } k > 0 \end{cases} \qquad (4.16)$$

Next we consider the inverse trigonometric functions. We have

$$\frac{d}{dx}\tan^{-1}g(x) = \frac{g'(x)}{1 + (g(x))^2}$$

The coefficients of the denominator function $1 + g^2$ must be formed; this is easily accomplished as $1 + g \cdot g$. Using again relation 4.7, we get

$$(\tan^{-1}g)_k = \begin{cases} \tan^{-1}g_0 & \text{if } k = 0 \\ \dfrac{1}{(1 + g^2)_0}\left[g_k - \dfrac{1}{k}\sum_{i=1}^{k-1} i(\tan^{-1}g)_i(1 + g^2)_{k-i}\right] & \text{if } k > 0 \end{cases} \tag{4.17}$$

In similar fashion, from

$$\frac{d}{dx}\sin^{-1}g(x) = \frac{g'(x)}{\sqrt{1 - (g(x))^2}}$$

we obtain

$$(\sin^{-1}g)_k = \begin{cases} \sin^{-1}g_0 & \text{if } k = 0 \\ \dfrac{1}{(\sqrt{1 - g^2})_0}\left[g_k - \dfrac{1}{k}\sum_{i=1}^{k-1} i(\sin^{-1}g)_i(\sqrt{1 - g^2})_{k-i}\right] & \text{if } k > 0 \end{cases} \tag{4.18}$$

The $\cos^{-1}$ function can be evaluated by using the relation $\cos^{-1}x = \pi/2 - \sin^{-1}x$.

For the max and min functions we have

$$(\max g, h)_k = \begin{cases} k = 0 & : \quad \max(g_0, h_0) \\ k > 0 & : \quad \begin{cases} g_k & \text{if } g_0 > h_0 \\ h_k & \text{if } g_0 \lessdot h_0 \\ \text{Error} & \text{if } g_0 \doteq h_0 \end{cases} \end{cases} \tag{4.19}$$

$$(\min g, h)_k = \begin{cases} k = 0 & : \quad \min(g_0, h_0) \\ k > 0 & : \quad \begin{cases} g_k & \text{if } g_0 < h_0 \\ h_k & \text{if } g_0 > h_0 \\ \text{Error} & \text{if } g_0 \doteq h_0 \end{cases} \end{cases}$$

This completes the description of the process for computing power series of an elementary function $f(x)$. Nonsolvable Problem 4.1 now can be made solvable by restricting $f(x)$ to these functions.

SOLVABLE PROBLEM 4.2: Given an elementary function $f(x)$ and a point x_0 at which $f(x)$ is infinitely differentiable, find to k decimals any derivative of $f(x)$ at x_0.

An elementary function's evaluation list makes it possible to obtain a power series for the function at the series expansion point x_0, and the power series coefficients yield the function's derivatives at x_0.

As an example of the general procedure, suppose we want the derivatives of the simple function $f(x) = x^2 + 5$ at the point $x = 3$. The evaluation list has two operations, a multiplication of x by itself and then the addition of a series representing the constant 5. The series expansion point is at $x = 3$, so first the primitive x is set to the form

$$x = 3 + 1(x - 3)\bullet$$

Here we use the $\bullet$ symbol to indicate that all higher power terms in $(x - 3)$ have zero coefficients. The execution of the first evaluation operation gives the result

$$9 + 6(x - 3) + 1(x - 3)^2\bullet$$

and the execution of the second gives us the power series for $f(x)$:

$$14 + 6(x - 3) + 1(x - 3)^2\bullet$$

From the series coefficients we find $f(3) = 14$, $f'(3) = 1! \cdot 6$, $f''(3) = 2! \cdot 1$, and all higher derivatives are zero.

4.4. COMPUTING THE DEFINITE INTEGRAL OF AN ELEMENTARY FUNCTION $f(x)$

It will be the purpose of this section to introduce a method of handling

SOLVABLE PROBLEM 4.3: Given an elementary function $f(x)$ defined in $[a, b]$, and with $f(x)$ infinitely differentiable in (a, b), with the exception, perhaps, of a finite number of points, find to k decimals the definite integral $\int_a^b f(x)\,dx$.

The method for finding definite integrals employs power series expansions. The basic idea is the following. Suppose we want to approximate the definite integral of $f(x)$ over an interval $[x_0 - h, x_0 + h]$. We form the series expansion for $f(x)$ up to the term in $(x - x_0)^n$, using x_0, the midpoint of our interval, as the series expansion point, and choosing the integer n reasonably large, say 12 or higher. We obtain a polynomial

$$f_0 + f_1(x - x_0) + f_2(x - x_0)^2 + \cdots + f_n(x - x_0)^n \tag{4.20}$$

which we integrate to obtain an approximation to $\int_{x_0-h}^{x_0+h} f(x)\,dx$.

A bound on the error made in replacing $f(x)$ by the Taylor polynomial must be computed. If the series expansion point x_0 were to vary in the interval $x_0 \pm h$, all coefficients f_i of the $f(x)$ power series would likewise vary. In the next section we show that if we can bound the variation of the last coefficient f_n in an interval $f_n \pm W$, then for any point x in $x_0 \pm h$, we have the following interval expression for $f(x)$:

$$f(x) = f_0 + f_1(x - x_0) + \cdots + f_{n-1}(x - x_0)^{n-1} \qquad (4.21)$$
$$+ (f_n \pm W)(x - x_0)^n$$

Now with an error bound, we are ready to integrate over our interval. It is convenient here to choose n equal to an even integer $2q$. We have then

$$\int_{x_0-h}^{x_0+h} f(x) \, dx = \int_{x_0-h}^{x_0+h} \sum_{i=0}^{2q} f_i(x - x_0)^i \, dx \pm \int_{x_0-h}^{x_0+h} W(x - x_0)^{2q} \, dx \qquad (4.22)$$

$$= 2\left[f_0 h + f_2 \frac{h^3}{3} + f_4 \frac{h^5}{5} + \cdots + f_{2q} \frac{h^{2q+1}}{2q + 1} \right] \pm 2W \frac{h^{2q+1}}{2q + 1}$$

The general plan is to divide the interval $[a, b]$ into sufficiently small subintervals, and in each subinterval use the expression 4.22 to replace the integral. We need some way of determining the bound W in each of these subintervals, and after we prove 4.21 we consider this problem.

4.5. TAYLOR'S FORMULA WITH INTEGRAL REMAINDER

Using the integration by parts formula, we obtain for the integral $\int_{x_0}^{x} f^{(n)}(t) \dfrac{(x - t)^{n-1}}{(n-1)!} \, dt$ the expression

$$\int_{x_0}^{x} f^{(n)}(t) \frac{(x - t)^{n-1}}{(n - 1)!} \, dt = f^{(n-1)}(t) \frac{(x - t)^{n-1}}{(n - 1)!} \Bigg|_{x_0}^{x} + \int_{x_0}^{x} f^{(n-1)}(t) \frac{(x - t)^{n-2}}{(n - 2)!} \, dt$$

$$= -f^{(n-1)}(x_0) \frac{(x - x_0)^{n-1}}{(n - 1)!} + \int_{x_0}^{x} f^{(n-1)}(t) \frac{(x - t)^{n-2}}{(n - 2)!} \, dt$$

If we keep using integration by parts on the right-hand integral, each time obtaining a term $-f^{(k)} \dfrac{(x - x_0)^k}{k!}$, eventually we obtain the integral $\int_{x_0}^{x} f'(t) \, dt$ which equals $f(x) - f(x_0)$. We can express our result as follows:

$$f(x) = f(x_0) + \frac{f'(x_0)}{1!} (x - x_0) + \cdots + \frac{f^{(n-1)}(x_0)}{(n - 1)!} (x - x_0)^{n-1}$$

$$+ \int_{x_0}^{x} f^{(n)}(t) \frac{(x - t)^{n-1}}{(n - 1)!} \, dt$$

This is a version of Taylor's formula with integral remainder. Note that

$$\int_{x_0}^{x} f^{(n)}(x_0) \frac{(x-t)^{n-1}}{(n-1)!} \, dt = f^{(n)}(x_0) \int_{x_0}^{x} \frac{(x-t)^{n-1}}{(n-1)!} \, dt = f^{(n)}(x_0) \frac{(x-x_0)^n}{n!}$$

Therefore $f(x)$ equals

$$f(x_0) + \frac{f'(x_0)}{1!}(x-x_0) + \cdots + \frac{f^{(n-1)}(x_0)}{(n-1)!}(x-x_0)^{n-1} + \frac{f^{(n)}(x_0)}{n!}(x-x_0)^n$$

with a remainder equal to

$$\int_{x_0}^{x} (f^{(n)}(t) - f^{(n)}(x_0)) \frac{(x-t)^{n-1}}{(n-1)!} \, dt$$

If w is an upper bound for $|f^{(n)}(t) - f^{(n)}(x_0)|$ over the interval from x_0 to x, then the magnitude of the remainder is not larger than

$$\int_{x_0}^{x} \frac{w|(x-t)^{n-1}|}{(n-1)!} \, dt = \frac{w|x-x_0|^n}{n!}$$

Our result may be written in interval form as

$$f(x) = f(x_0) + \frac{f'(x_0)}{1!}(x-x_0) + \cdots + \frac{f^{(n-1)}(x_0)}{(n-1)!}(x-x_0)^{n-1}$$
$$+ \frac{f^{(n)}(x_0) \pm w}{n!}(x-x_0)^n \tag{4.23}$$

After substituting f_i for $f^{(i)}(x_0)/i!$ consistently, and replacing $w/n!$ by W, we obtain Equation 4.21.

4.6. FORMAL INTERVAL ARITHMETIC

In Chapter 2, we introduced interval arithmetic as a means of monitoring the computation error of range arithmetic. We can use interval arithmetic again in a different fashion to obtain the needed bounds on $f^{(2q)}(x)$. The general idea here is to do the computation for the $f(x)$ power series in the way previously described, except that an interval $m \pm w$ is substituted for each series coefficient appearing in the computation, and interval arithmetic is used everywhere. It is true that range arithmetic itself is a form of interval arithmetic, but a ranged number has only a gross representation of the error, mainly as a means of automatically discarding mantissa digits as they become suspect. In this application we want a more accurate representation of the error, or rather the "wid." So one ranged number is used for the mid m, and one for the wid w. After a computation is complete, as usual we can check our results to determine whether we have obtained enough mantissa digits, and if not, the whole computation is repeated at a higher precision.

We repeat below the rules of interval arithmetic. The division operation has a slightly different form than previously presented in that dotted relations are used. This is because now the two quantities defining an interval are both ranged numbers, not rational numbers.

$$m_1 \pm w_1 \quad + \quad m_2 \pm w_2 = (m_1 + m_2) \pm (w_1 + w_2) \tag{4.24}$$

$$m_1 \pm w_1 \quad - \quad m_2 \pm w_2 = (m_1 - m_2) \pm (w_1 + w_2) \tag{4.25}$$

$$m_1 \pm w_1 \quad \times \quad m_2 \pm w_2 = (m_1 m_2) \pm (w_1 |m_2| + |m_1| w_2 + w_1 w_2) \tag{4.26}$$

$$m_1 \pm w_1 \quad \div \quad m_2 \pm w_2 = \begin{cases} \left(\dfrac{m_1}{m_2}\right) \pm \left(\dfrac{w_1 + \left|\frac{m_1}{m_2}\right| w_2}{|m_2| - w_2}\right) & \text{if } |m_2| > w_2 \\ \text{Division error} & \text{if } |m_2| \leq w_2 \end{cases} \tag{4.27}$$

In the computation for the $f(x)$ power series over an interval $[x_0 - h, x_0 + h]$, constants and the variable x, which are the primitives, now have the form:

$$c = (c \pm 0)\bullet$$

$$x = (x_0 \pm h) + (1 \pm 0)(x - x_0)\bullet$$

All the steps of the evaluation list for $f(x)$ now are done using the interval arithmetic relations above, and we call such computation *formal interval arithmetic* (range arithmetic being *informal*). Instead of obtaining for $f(x)$ the series 4.20, we now obtain for $f(x)$ the representation

$$(f_0 \pm w_0) + (f_1 \pm w_1)(x - x_0) + \cdots + (f_n \pm w_n)(x - x_0)^n \tag{4.28}$$

Referring to equation (4.22), we see that with n set equal to the even integer $2q$, the expression for the integral is

$$\int_{x_0-h}^{x_0+h} f(x)\,dx = 2\left[f_0 h + \frac{f_2 h^3}{3} + \frac{f_4 h^5}{5} + \cdots + \frac{f_{2q} h^{2q+1}}{2q + 1} \right]$$
$$\pm 2w_{2q} \frac{h^{2q+1}}{2q + 1} \tag{4.29}$$

We use only the wid w_{2q}, but to obtain this wid we must compute wids for all coefficients of all series that arise in generating the $f(x)$ series.

Because the various series operations involve function evaluations, these need to be done in interval arithmetic style. Generally for each standard function it is not difficult to obtain an appropriate interval arithmetic relation. As an example, consider the natural logarithm function. Because $\ln x$ is an increasing function with a decreasing derivative, we have

$$\ln(m \pm w) = \ln m \pm (\ln m - \ln(m - w))$$

Two logarithm evaluations are required here, and each evaluation is likely to require many arithmetic operations. To speed up logarithm evaluation, one might prefer to use the less accurate relation

$$\ln(m \pm w) = \ln m \pm \frac{w}{m - w}$$

obtained by applying the mean value theorem: $\ln m - \ln(m - w) = (\ln' c)w = (1/c)\cdot w \leq w/(m-w)$. Thus with the standard functions, there are often several choices for the interval arithmetic evaluation equations. We list below some of the relations used in our C++ software for standard function evaluation.

$$e^{m \pm w} = e^m \pm e^m(e^w - 1) \tag{4.30}$$

$$\ln(m \pm w) = \begin{cases} \ln m \pm \dfrac{w}{m - w} & \text{if } m > w \\ \ln \text{error} & \text{if } m \leq w \end{cases} \tag{4.31}$$

$$\sin(m \pm w) = \sin m \pm \min(|\cos m| + w, 1) \cdot w \tag{4.32}$$

$$\cos(m \pm w) = \cos m \pm \min(|\sin m| + w, 1) \cdot w \tag{4.33}$$

$$\sinh(m \pm w) = \sinh m \pm \cosh m \cdot (e^w - 1) \tag{4.34}$$

$$\cosh(m \pm w) = \cosh m \pm \cosh m \cdot (e^w - 1) \tag{4.35}$$

$$\tan^{-1}(m \pm w) = \tan^{-1}m \pm \frac{w}{1 + [\max(|m| - w, 0)]^2} \tag{4.36}$$

4.7. USING A TASK QUEUE TO COMPUTE A DEFINITE INTEGRAL

In finding an approximation to $\int_a^b f(x)dx$ along with its error bound, it is convenient to arrange the computation by means of a *task queue*. A task queue often is a convenient way of handling computations where the exact sequence of steps needed to perform the computation is not known in advance. A task queue finds many applications in precise computation programs. In general a task queue is a list of the tasks that must be performed to reach a desired objective. Depending on the problem, the list initially holds a single task, or a short list of tasks, which would attain the objective, but generally it is not known whether these tasks can be performed successfully. Cyclically, the task at the head of the list is attempted, discarded from the list if the task is completed successfully, and this continues until the task queue is empty, the objective having been attained. If the task at the head of the list cannot be done, the task is split into an appropriate set of simpler subtasks that accomplish the same end, and these subtasks replace the task at the head of the list.

When dealing with definite integrals it is convenient to always have [0, 1] as the integration interval. If we make the substitution $x = a + (b - a)u$, the definite integral $\int_a^b f(x)dx$ can be rewritten as $\int_0^1 F(u)du$, where $F(u)$ equals $(b - a) f (a + (b - a)u)$. If our integral is required to k correct fixed-place decimal places, then we must obtain an integral approximation with the error bound ϵ equal to ϵ_k of Equation 3.1. The task queue holds a list of subintervals $u_0 \pm h$ on each of which the integral of $F(u)$ must be found. Initially the list holds the single interval 0.5 ± 0.5. A task cycle consists of calculating the power series 4.28 and integral 4.29 for the first queue interval $u_0 \pm h$, and checking that the error bound $2w_{2q} h^{2q+1}/(2q+1)$ does not exceed its ϵ allotment, which is $\epsilon \cdot$ (length $[u_0-h, u_0+h]$/length $[0, 1]$) or $2h\epsilon$. If this requirement is met, the integral approximation is added to a sum S, the error bound is added to another sum E, both sums initially being zero, and the queue interval is discarded. Otherwise, the queue interval is bisected and the two subintervals replace it on the task queue. Because the error of an integral approximation has the multiplier h^{2q+1}, when the width h is halved, the error bound decreases by at least the factor $1/2^{2q+1}$ while the ϵ allotment decreases only by the factor $\frac{1}{2}$. Eventually, when h gets small enough, the error bound test is passed. When finally the task queue is empty, S holds the required approximation to $\int_0^1 F(u) \, du$ and E holds the error bound, which does not exceed ϵ.

If the integral error bound for a queue interval is more than its allotment $2h\epsilon$, it is advisable to test whether the precision of computation is adequate and to increase it if not. Such a test is needed at some point in any range arithmetic program.

It may happen that the $F(u)$ series cannot be computed for a queue interval because the interval is too large. Such a series generation error may occur, for instance, on a division operation $g(u)/h(u)$, when the leading term of $h(u)$ is an interval containing the zero point. Or a series error may occur when generating $\ln g(u)$ if the leading term of g contains negative numbers in its interval. These cases are treated just as if the ϵ allotment were exceeded.

The procedure so far described is satisfactory as long as the function $F(u)$ is analytic at every point of [0,1]. (A function is *analytic* at a point u_0 if it has derivatives of arbitrary order there.) But of course it is possible that $F(u)$ is defined but is not analytic at a few points inside [0, 1] or at an endpoint. For instance the integral might be $\int_0^1 \sqrt{1 - u^3} \, du$ with the nonanalytic point $u = 1$, or $\int_0^1 (u - \frac{1}{2})^{1/3} \, du$ with the nonanalytic point $u = \frac{1}{2}$. It will not be possible to obtain any series coefficients beyond the constant coefficient when the integration interval $u_0 \pm h$ contains such a point. This amounts to doing only an interval function evaluation over $u_0 \pm h$. We obtain then just the initial series term $f_0 \pm w_0$.

However, Equation 4.29 still applies, and we obtain for the integral the value $2f_0h$ and the error bound $2w_0h$. The wid w_0 decreases as h decreases, so the error $2w_0h$ eventually becomes less than the error allotment $2\epsilon h$ when h becomes small enough. Normally an interval $u_0 \pm h$ containing a nonanalytic point is extremely small before it passes this test.

4.8. COMPUTING IMPROPER INTEGRALS

In this section we consider what can be done in the way of accurately computing various improper integrals. Improper integrals can be of the form $\int_a^b f(x)\,dx$ where the function $f(x)$ is not defined throughout $[a, b]$ but only in $(a, b]$, only in $[a, b)$, or only in (a,b). Another common improper integral is $\int_a^\infty f(x)\,dx$. Here $f(x)$ may be defined in $[a, \infty)$ or possibly only in (a, ∞). A similar improper integral is $\int_{-\infty}^b f(x)\,dx$. Finally there is the improper integral $\int_{-\infty}^\infty f(x)\,dx$.

Consider first the integral $\int_a^b f(x)\,dx$. It can not be the task of a program to determine from the supplied function $f(x)$ and supplied constants a and b *whether or not* the integral converges. The integral $\int_0^1 x^a\,dx$ converges if $a > -1$ and diverges otherwise, so a program that could decide on convergence would be able to decide whether or not a was greater than -1, contradicting Nonsolvable Problem 3.2. Similarly, for programs handling convergent integrals only, it can not be their task to determine the integral to k places from just the supplied function $f(x)$ and the constants a and b. Additional information must be supplied. To see this, consider the integral $\int_0^1 f_a(x)\,dx$ where $f_a(x)$ is defined for positive x by the equation below for any value of the parameter a.

$$f_a(x) = \begin{cases} \max\left(\dfrac{1}{a^2}(a - x), 0\right) & \text{if } 0 < a \\ 0 & \text{if } a \le 0 \end{cases}$$

When $0 < a \le 1$, we have $\int_0^1 f_a(x)\,dx = \frac{1}{2}$, the area of the spike of $f_a(x)$ always equalling $\frac{1}{2}$ (Fig. 4.1). When $a \le 0$, we have $\int_0^1 f_a(x)\,dx = 0$. A program that could return the value of the integral could be used to decide whether or not the parameter a was positive, which amounts to attempting Nonsolvable Problem 3.2. Note there is no difficulty computing $f_a(x)$ to arbitrary accuracy for any positive x by the following procedure. Because x is known to be positive, we first obtain a ranged approximation to x such that its left endpoint is positive, say δ. Then we obtain an a approximation with wid less than $\frac{\delta}{2}$. If $a < x$, then $f_a(x) = 0$. If $a \ge x$, then $a > 0$, so we can evaluate $f_a(x)$ to any required accuracy using the first line of the f_a definition.

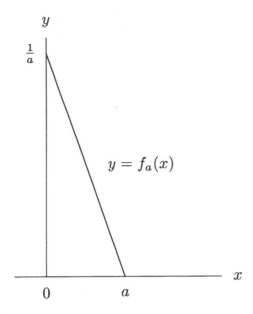

FIGURE 4.1   The Function $f_a(x)$ for Positive a.

A possible solvable problem is

SOLVABLE PROBLEM 4.4: Given a convergent improper integral $\int_a^b f(x)\,dx$, with the elementary function $f(x)$ defined in the interval $(a, b]$, and with an elementary function $g(x)$ defined for x in some interval $(a, b_1]$ in which the relations $\left| \int_a^x f(t)\,dt \right| \le g(x)$ and $\lim\limits_{x \to a^+} g(x) = 0$ are true, find $\int_a^b f(x)$ to k decimals.

Let ϵ be an error bound, which if achieved for an integral approximation, allows us to display the integral to the desired number of decimal places. (Using equation [3.1], for k fixed decimals places, we can take ϵ as ϵ_k, and for k floating decimals, we can take ϵ as ϵ_{2k}.) We compute $g(\alpha_n)$, where $\alpha_n = a + \frac{b_1 - a}{2^n}$, for cases $n = 0, 1, 2, \ldots$, until finally we obtain a g value less than $\epsilon/2$. If this occurs for the argument α_N, then we compute the definite integral $\int_{\alpha_N}^b f(x)dx$ to the error bound $\epsilon/2$ by the method described in the preceding section.

Usually it is not difficult to find a suitable function $g(x)$. For example, consider the integral $\int_0^1 \frac{\cos x}{\sqrt{x}}\,dx$. We have

$$\left| \int_0^x \frac{\cos t}{\sqrt{t}}\,dt \right| \le \int_0^x \frac{dt}{\sqrt{t}} = 2\sqrt{x}$$

so $g(x)$ can be taken to be $2\sqrt{x}$, with $(0, 1]$ being its domain.

For improper integrals $\int_a^\infty f(x)\,dx$, a suitable problem is

SOLVABLE PROBLEM 4.5: Given a convergent improper integral $\int_a^\infty f(x) \, dx$, with the elementary function $f(x)$ defined over $[a, \infty)$ and an elementary function $h(x)$ defined for x in some interval $[a_1, \infty)$ in which the relations $\left| \int_x^\infty f(t) \, dt \right| \leq h(x)$ and $\lim\limits_{x \to \infty} h(x) = 0$ are true, find $\int_a^\infty f(x) \, dx$ to k decimals.

Again let ϵ be an error bound, which if achieved for an integral approximation, allows us to display the integral to the desired number of places. We compute $h(\beta_n)$, where $\beta_n = \max(a, a_1, 1) \cdot 1.1^{n-1}$, for the cases $n = 1, 2, \ldots$, until we obtain an h value less than $\epsilon/2$. If this occurs for the argument β_M, then we compute the definite integral $\int_a^{\beta_M} f(x) \, dx$ to the error bound $\epsilon/2$ by the method described in the preceding section.

Thus for the improper integral $\int_0^\infty \sqrt{x} e^{-x^2} \, dx$, for x in $[1, \infty)$ we have

$$\left| \int_x^\infty \sqrt{t} e^{-t^2} \, dt \right| \leq \int_x^\infty t e^{-t^2} \, dt = \frac{1}{2} e^{-x^2}$$

so we can take $h(x)$ equal to $\frac{1}{2} e^{-x^2}$, with the domain $[1, \infty)$.

For some integrals we may need to find both a function $g(x)$ and a function $h(x)$.

SOLVABLE PROBLEM 4.6: Given the convergent improper integral $\int_a^\infty f(x) \, dx$, with the elementary function $f(x)$ defined over (a, ∞), an elementary function $g(x)$ defined for x in some interval $(a, b_1]$ in which the relations $\left| \int_a^x f(t) \, dt \right| \leq g(x)$ and $\lim\limits_{x \to a^+} g(x) = 0$ are true, and an elementary function $h(x)$ defined for x in some interval $[a_1, \infty)$ in which the relations $\left| \int_x^\infty f(t) \, dt \right| \leq h(x)$ and $\lim\limits_{x \to \infty} h(x) = 0$ hold, find $\int_a^\infty f(x) \, dx$ to k decimals.

The error $\epsilon/4$ is allotted for each function $g(x)$ or $h(x)$, in the way previously described, and the definite integral $\int_{\alpha_N}^{\beta_M} f(x) \, dx$ is computed to the error bound $\epsilon/2$.

An example is $\int_0^\infty e^{-(x+1/x)} \, dx$. For x in $(0, 0.01]$, we have

$$\left| \int_0^x e^{-(t+1/t)} \, dt \right| < xe^{-100}$$

and for x in $[1, \infty)$, we have

$$\left| \int_x^\infty e^{-(t+1/t)} \, dt \right| < \int_x^\infty e^{-t} \, dt = e^{-x}$$

Thus $g(x)$ can be taken as xe^{-100} with the domain $(0, 0.01]$, and $h(x)$ as e^{-x} with the domain $[1, \infty)$.

A program using this computation method for various improper integral types is included among the demonstration programs.

4.9. POWER SERIES FOR ELEMENTARY FUNCTIONS OF SEVERAL VARIABLES

The equations developed in earlier sections allow the generation of a power series for an elementary function $f(x)$ and the bounding of the series truncation error. Here we obtain similar equations for an elementary function of several variables $f(x_1, \ldots, x_n)$. If the series expansion point is $(x_{1_0}, \ldots, x_{n_0})$, and if the function f has all partial derivatives with respect to its variables in the region R consisting of the product of intervals $x_{i_0} \pm h_i$, $i = 1, \ldots, n$, then f can be expressed as a power series of the form

$$\sum_{d=0}^{\infty} \sum_{k_1 + \cdots + k_n = d} f_{k_1, \ldots, k_n} (x_1 - x_{1_0})^{k_1} \cdots (x_n - x_{n_0})^{k_n} \qquad (4.37)$$

The power series term

$$f_{k_1, \ldots, k_n} (x_1 - k_{1_0})^{k_1} \cdots (x_n - x_{n_0})^{k_n}$$

is said to be of degree d, where d equals the sum $k_1 + \ldots + k_n$. If all series terms up to degree D are obtained, a bound on the truncation error can be obtained by adapting the Taylor series formula with remainder given in Section 4.5. If $(x_1, \ldots, x_n)$ is some point in R, define the function $g(t)$ by the equation

$$g(t) = f(x_{1_0} + t(x_1 - x_{1_0}), \ldots, x_{n_0} + t(x_n - x_{n_0}))$$

The variable t has the domain $[0, 1]$, so that $g(0) = f(x_{1_0}, \ldots, x_{n_0})$, and $g(1) = f(x_1, \ldots, x_n)$. Expanding $g(t)$ in a power series about the point $t = 0$, we have

$$g(t) = g(0) + \frac{g'(0)}{1!} t + \frac{g''(0)}{2!} t^2 + \cdots + \frac{g^{(D-1)}(0)}{(D-1)!} t^{D-1}$$

$$+ \frac{g^{(D)}(0) \pm w}{D!} t^D \qquad (4.38)$$

where w bounds the variation of $g^{(D)}$ from $g^{(D)}(0)$ as t varies in $[0, 1]$. Because dg/dt at the point $t = 0$ equals $\sum_{i=1}^{n} (x_i - x_{i_0}) \, \partial f/\partial x_i$ with all partial derivatives taken at the point $(x_{i_0}, \ldots, x_{n_0})$, the coefficient $g^{(k)}(0)/k!$ equals

$$\frac{1}{k!} \left[(x_1 - x_{1_0}) \frac{\partial}{\partial x_1} + \cdots + (x_n - x_{n_0}) \frac{\partial}{\partial x_n} \right]^k f(x_{1_0}, \ldots, x_{n_0})$$

After the multinomial is expanded, this is

$$\sum_{k_1+\cdots+k_n=k} \frac{1}{k_1! \cdots k_n!} \left[\frac{\partial^k}{\partial x_1^{k_1} \cdots \partial x_n^{k_n}} f(x_{10}, \ldots, x_{n0}) \right] \cdot$$

$$\cdot (x_1 - x_{10})^{k_1} \cdots (x_n - x_{n0})^{k_n} \qquad (4.39)$$

Thus the coefficient $f_{k_1,\ldots,k_n}$ of equation (4.37) equals

$$\frac{1}{k_1! \cdots k_n!} \cdot \frac{\partial^k}{\partial x_1^{k_1} \cdots \partial x_n^{k_n}} f(x_{10}, \ldots, x_{n0})$$

In equation (4.38) when we set t equal to 1, we obtain

$$f(x_1, \ldots, x_n) = \sum_{d=0}^{D-1} \sum_{k_1+\cdots+k_n=d} f_{k_1,\ldots,k_n}(x_1 - x_{10})^{k_1} \cdots (x_n - x_{n0})^{k_n} + \qquad (4.40)$$

$$\sum_{k_1+\cdots+k_n=D} (f_{k_1,\ldots,k_n} \pm w_{k_1,\ldots,k_n})(x_1 - x_{10})^{k_1} \cdots (x_n - x_{n0})^{k_n}$$

In obtaining the relations for generating the terms of an elementary function $f(x)$ in Section 4.3, often a derivative with respect to x was taken, followed by the multiplication of the resulting equation by $(x - x_0)$. The two steps can be performed together by using the operator $(x - x_0)\frac{d}{dx}$. For a function of the variables $x_1, \ldots, x_n$, the corresponding operator is

$$(x_1 - x_{10}) \frac{\partial}{\partial x_1} + \cdots + (x_n - x_{n0}) \frac{\partial}{\partial x_n} \qquad (4.41)$$

Using this operator, we can derive equations for the power series terms of functions with an arbitrary number of variables. The steps are similar to those taken in the Section 4.3.

Some additional notation is helpful here. We use the capital letter K to represent the index $k_1, \ldots, k_n$, so the coefficient $g_{k_1, \ldots, k_n}$ of the series term

$$g_{k_1, \ldots, k_n}(x_1 - x_{10})^{k_1} \cdots (x_n - x_{n0})^{k_n}$$

becomes g_K. The capital letter O is reserved for the index $0, \ldots, 0$. Define $[K]$ to be the integer $k_1 + \ldots + k_n$, the degree of the term g_K. For two indices I, J, define $I - J$ to be the index Q with $q_1 = i_1 - j_1, \ldots, q_n = i_n - j_n$. The notation $I = J$ has the obvious interpretation, and $I \leq J$ signifies

$$i_1 \leq j_1, \ldots, i_n \leq j_n \qquad (4.42)$$

The notation $I < J$ also signifies (4.42) but with the requirement that inequality occurs at least once. Thus if $[I] = [J]$, which implies that g_I and g_J have the same degree, then neither $I < J$ nor $I > J$ is possible. Using the operator (4.41), the following equations can be derived by repeating the steps of Section 4.3:

$$(g + h)_K = g_K + h_K$$

$$(g - h)_K = g_K - h_K$$

$$(g \cdot h)_K = \sum_{0=I \leq K} g_I h_{K-I} = \sum_{0=I \leq K} g_{K-I} - h_I$$

$$(g/h)_K = \begin{cases} g_0/h_0 & \text{if } K = 0 \\ \dfrac{1}{h_0}\left(g_K - \sum_{0=I<K} (g/h)_I h_{K-I} \right) & \text{if } K > 0 \end{cases}$$

$$(e^g)_K = \begin{cases} e^{g_0} & \text{if } K = 0 \\ \dfrac{1}{[K]} \sum_{0<I \leq K} [I] g_I (e^g)_{K-I} & \text{if } K > 0 \end{cases}$$

$$(\ln g)_K = \begin{cases} \ln g_0 & \text{if } K = 0 \\ \dfrac{1}{g_0}\left(g_K - \dfrac{1}{[K]} \sum_{0<I<K} [I](\ln g)_I g_{K-I} \right) & \text{if } K > 0 \end{cases}$$

$$(\sin g)_K = \begin{cases} \sin g_0 & \text{if } K = 0 \\ \dfrac{1}{[K]} \sum_{0<I \leq K} [I] g_I (\cos g)_{K-I} & \text{if } K > 0 \end{cases}$$

$$(\cos g)_K = \begin{cases} \cos g_0 & \text{if } K = 0 \\ -\dfrac{1}{[K]} \sum_{0<I \leq K} [I] g_I (\sin g)_{K-I} & \text{if } K > 0 \end{cases}$$

$$(\sinh g)_K = \begin{cases} \sinh g_0 & \text{if } K = 0 \\ \dfrac{1}{[K]} \sum_{0<I \leq K} [I] g_I (\cosh g)_{K-I} & \text{if } K > 0 \end{cases}$$

$$(\cosh g)_K = \begin{cases} \cosh g_0 & \text{if } K = 0 \\ \dfrac{1}{[K]} \sum_{0<I \leq K} [I] g_I (\sinh g)_{K-I} & \text{if } K > 0 \end{cases}$$

$$(g^a)_K = \begin{cases} g_0^a & \text{if } K = 0 \\ \dfrac{1}{g_0} \sum_{0<I \leq K} \left(\dfrac{(a+1)[I]}{[K]} - 1 \right)(g^a)_{K-I} g_I & \text{if } K > 0 \end{cases}$$

$$(g^a)_K = \begin{cases} g_0^{a-1} g_0 & \text{if } K = 0 \\ \dfrac{a}{[K]} \sum_{0<I \leq K} (g^{a-1})_{K-I} [I] g_I & \text{if } K > 0 \end{cases}$$

$$(\tan^{-1}g)_K = \begin{cases} \tan^{-1}g_0 & \text{if } K = O \\ \dfrac{1}{(1+g^2)_0}\left[g_K - \dfrac{1}{[K]}\sum_{O<I<K}[I](\tan^{-1}g)_I(1+g^2)_{K-I}\right] & \text{if } K > O \end{cases}$$

$$(\sin^{-1}g)_K = \begin{cases} \sin^{-1}g_0 & \text{if } K = O \\ \dfrac{1}{(\sqrt{1-g^2})_0}\left[g_K - \dfrac{1}{[K]}\sum_{O<I<K}[I](\sin^{-1}g)_I(\sqrt{1-g^2})_{K-I}\right] & \text{if } K > O \end{cases}$$

$$(\max\, g,\, h)_K = \begin{cases} K = O &:& \max(g_0, h_0) \\ K > O &:& \begin{cases} g_K & \text{if } g_0 > h_0 \\ h_K & \text{if } g_0 < h_0 \\ \text{Error} & \text{if } g_0 \doteq h_0 \end{cases} \end{cases}$$

$$(\min\, g,\, h)_K = \begin{cases} K = O &:& \min(g_0, h_0) \\ K > O &:& \begin{cases} g_K & \text{if } g_0 < h_0 \\ h_K & \text{if } g_0 > h_0 \\ \text{Error} & \text{if } g_0 \doteq h_0 \end{cases} \end{cases}$$

These formulas allow the generation of the power series for a function of any number of variables. The degree 0 term of a series is generated first, then the degree 1 terms, and so on up to the terms of highest degree D. After the series expansion point $(x_{1_0}, \ldots x_{n_0})$ is chosen, the primitives are

$$c = c\bullet$$

for any constant c, and

$$x_i = x_{i_0} + 1(x_i - x_{i_0})\bullet$$

for any variable x_i.

4.10. COMPUTING HIGHER DIMENSIONAL INTEGRALS

Consider the iterated integral $\int_a^b \int_{g_1(x)}^{g_2(x)} f(x, y)\, dx\, dy$ in which the region of integration defined by the limit functions $g_2(x)$ and $g_1(x)$ over their domain $[a,b]$ is a bounded region R. The problem of finding such an integral to k decimals can be done by a procedure similar to that suggested for the definite integral $\int_a^b f(x)dx$. Here the conditions analogous to the conditions of Solvable Problem 4.3 are that the elementary function f be defined in R, and analytic there except for a finite number of curves in R, and that the elementary functions g_1 and g_2 be defined in $[a, b]$ and analytic there except for a finite number of points. However, such computations usually become computer bound unless the function f is analytic throughout R, and the functions g_1 and g_2 are analytic in $[a, b]$.

First the integration is converted to integration over the unit square by making the substitutions

$$y = g_1(x) + (g_2(x) - g_1(x))v \qquad (4.43)$$

$$x = a + (b - a)u$$

with u and v both varying in $[0, 1]$. The integral then has the form $\int_0^1 \int_0^1 F(u, v)\,du\,dv$ where

$$F(u, v) = (b - a)(g_2(x) - g_1(x))f(x, y)$$

with x and y changed to functions of u and v using (4.43). A task queue is used and holds a list of u, v subrectangles $(u_0 \pm h_1) \times (v_0 \pm h_2)$ over which integration needs to be done. Initially, the task queue holds just the square $(0.5 \pm 0.5) \times (0.5 \pm 0.5)$. The nonconstant primitives are

$$u = (u_0 \pm h_1) + (1 \pm 0)(u - u_0)\bullet$$

and

$$v = (v_0 \pm h_2) + (1 \pm 0)(v - v_0)\bullet$$

For the function $F(u, v)$ we obtain the degree $2q$ series

$$\sum_{d=0}^{2q} \sum_{i+j=d} (F_{i,j} \pm w_{i,j})(u - u_0)^i (v - v_0)^j$$

from which the integral approximation

$$\sum_{d=0}^{q} \sum_{s+t=d} \frac{F_{2s,2t} h_1^{2s+1} h_2^{2t+1}}{(2s + 1)(2t + 1)}$$

is obtained, and also the error bound

$$\sum_{i+j=2q} \frac{w_{i,j} h_1^{i+1} h_2^{j+1}}{(i + 1)(j + 1)}$$

If ϵ is the error bound that must be achieved by our final integral approximation to obtain from it k correct decimals, we need to check that our subrectangle error bound is not greater than the allotment $4h_1 h_2 \epsilon$.

When the error allotment is exceeded, the integration subrectangle is divided into two equal subrectangles, obtained by halving either the u dimension or the v dimension. The division method is chosen by recomputing the error bound with first h_1 replaced by $h_1/2$, then h_2 replaced by $h_2/2$, and letting the smaller value determine the division method. If there is a series generation error, then the larger side of the subrectangle is bisected.

A demonstration program computes multiple integrals, of second or higher order, and to prevent the computation from being computerbound, the program requires that the integrand and all limit functions of the integral be analytic within their domains.

EXERCISES

1. Call `deriv` and obtain the derivatives up to order 2 of the functions $\sin(x)$, $\sin(x + y)$, and $\sin(x + y + z)$ at the point where all variables are zero.

2. Any demonstration program should be capable of being halted at any time while it is running by entering a *control C* at the console keyboard. Verify that your system allows this interruption: Call `integ` and calculate $\int_{-1}^{1} x^{\frac{1}{3}} \, dx$ to 20 decimals. Interrupt this long computation by entering a *control C*.

3. Call `integ`, request five decimal places, and calculate $\int_0^{\pi} \sin x \, dx$. After `integ` terminates, the file `integ.log` has a record of all the keyboard entries. Now obtain a 10-decimal answer in the following way. Edit the file `integ.log` with an editor program, changing the number of deci mals from 5 to 10. Enter the command line `integ < integ.log` so that the input to `integ` is the file `integ.log` instead of console input, and note that a 10-decimal answer is obtained this time.

   All demo programs that create a `log` file also create a record of a successful run in a `prt` file for later viewing or printing. Thus after `integ` completes an integral calculation, in the file `integ.prt` there is a record of the computation, which can be viewed later or printed.

4. Call `integ` and calculate to 10 decimals $\int_{-1}^{1} \sqrt{1 - x^2} \, dx$. An integral is computed by summing integral approximations over small subintervals starting at the left interval endpoint and working toward the right endpoint. Note that progress over the interval $[-1, +1]$ is slow near the endpoints of the interval, where the derivative of the integrand gets large. At $x = -1$ and $x = 1$, the integrand derivative is not defined, and a degree 0 evaluation of the integrand is needed over small subintervals containing these points.

5. Call `impint` and calculate to five decimals the improper integral $\int_0^{\pi/2} \frac{\sin x}{x} \, dx$. The function $g(x)$ needed here can be taken as x because $\left|\frac{\sin x}{x}\right| < 1$ for x in $(0, \frac{\pi}{2}]$. Redo the calculation with $g(x)$ taken incorrectly as $x/1000$, and notice that the answer changes slightly from its previous correct value.

6. Call `impint` and obtain a five-decimal value for the integral $\int_1^2 \frac{\ln x}{\sqrt{x-1}}\, dx$. Note that by the Mean Value Theorem $\ln x - \ln 1 = \ln' c(x-1) < 1 \cdot (x-1)$, which makes it easy to find an appropriate function $g(x)$.

7. Call `impint` and obtain a five-decimal value for the integral $\int_{-\infty}^{\infty} \frac{e^{-x^2}}{1+x^2}\, dx$.

8. Call `integ_n` and evaluate to five decimals the integral $\int_1^2 dx \int_1^x x^2 e^{xy}\, dy$. Try also to compute to five decimals the integral $\int_0^1 dx \int_0^x \sqrt{1-y^2}\, dy$. This integral computation is rejected by `integ_n` because of difficulty obtaining a series expansion of the integrand near the (x, y) point $(1, 1)$. Note that the integrand is not analytic at $(1, 1)$.

NOTES

N4.1. R. Moore was an early advocate of power series methods for finding derivatives and integrals. We have used the notation of his book [4]. For a more complete discussion of derivatives and power series, see the book by Rall [5].

N4.2. The book by Davis and Rabinowitz [3] gives a comprehensive survey of other integration techniques.

N4.3. The papers by Corliss and Krenz [1] and Corliss and Rall [2] discuss alternate approaches to the accurate computation of definite integrals.

N4.4. In applied mathematics various functions that are not elementary are encountered often. Examples are the gamma and beta functions, the Bessel functions, and the hypergeometric functions. There is no basic difficulty in extending the elementary function class to include additional function types, as long as the new function types can be generated by employing specific power series formulas (as is the case for the standard functions). Such an extension preserves the ability to generate any defined derivative of an elementary function by using the function's evaluation list. A solvable problem listed in this text that specifies an elementary function, such as Problem 4.2 or 4.3, would still be valid for the extended class of elementary functions.

REFERENCES

1. Corliss, G. and Krenz, G., Indefinite integration with validation, *ACM Trans. Math. Software* **15** (1989), 375–393.

2. Corliss, G. and Rall, L. B., Adaptive, self-validating numerical quadrature, *SIAM J. Sci. Stat. Comput.* **8** (1987), 831–847.

3. Davis, P. J. and Rabinowitz, P., *Methods of Numerical Integration, 2nd Edit.,* Academic Press, New York, 1984.
4. Moore, R. E., *Methods and Applications of Interval Analysis,* SIAM Studies in Applied Mathematics, SIAM, Philadelphia, 1979, 24–29.
5. Rall, L. B., *Automatic Differentiation: Techniques and Applications,* Lecture Notes in Computer Science 120, Springer-Verlag, Berlin, 1981.

V

FINDING ZEROS OF REAL FUNCTIONS

5.I. THE GENERAL PROBLEM

In this chapter we consider the problem of finding where n elementary functions of n variables are simultaneously zero. The equations that must be solved are

$$f_1(x_1, x_2, \ldots, x_n) = 0$$
$$f_2(x_1, x_2, \ldots, x_n) = 0$$
$$\vdots \qquad\qquad \vdots$$
$$f_n(x_1, x_2, \ldots, x_n) = 0 \qquad (5.1)$$

An argument $(c_1, c_2, \ldots, c_n)$ at which all functions are zero is called a *zero* of the functions. In problems of this type there is some region of the argument space in which zeros are sought. We assume the region of interest is defined by restricting each variable to a finite interval:

$$a_i \leq x_i \leq b_i, \qquad i = 1, 2, \ldots, n$$

When n is 1, we have the simple problem of finding where in a given interval a single function of one variable is zero. First we consider how to properly specify this problem so it is solvable. From a solvable version for the case $n = 1$, we can hope to find a solvable version for the general case.

5.2. THE SINGLE FUNCTION PROBLEM

We assume that the elementary function $f(x)$ is defined in an interval $[a, b]$, and that all zeros inside this interval are desired. In general we would want a solution program to indicate that there are no zeros in $[a, b]$ when this is the case, and if there are zeros, to list them to a prescribed number of correct decimal places. We expect a program to accomplish this by examination of $f(x)$, and possibly the derivative $f'(x)$, at various arguments in $[a, b]$. The data that must be specified appear to be just the function f, the two numbers a and b defining the search interval, and the number k of correct decimal places wanted.

However, suppose for the search interval $[a, b]$ the supplied function is

$$f_c(x) = x - c$$

where the parameter c is constrained to a small interval centered at $x = a$. A difficulty here is that if we find $f_c(a) \doteq 0$, we are uncertain whether this means a zero lies in $[a, b]$. Perhaps the zero is just outside this interval. Our solution program must report a zero if $c \geq a$ and no zero otherwise, and this amounts to attempting Nonsolvable Problem 3.2. A similar problem could occur at the other endpoint b. We eliminate this difficulty by requiring, as part of the problem specification, that the function f be nonzero at both endpoints a and b.

We can decide if a zero is likely in $[a, b]$ by using formal interval arithmetic, as described in Chapter 4 for the problem of finding definite integrals, and compute an interval for $f(x)$ over $[a, b]$. We set the primitive x appropriately to reflect the interval $[a, b]$, and then by executing the $f(x)$ evaluation operations to degree 0, obtain an interval $m_0 \pm w_0$ for $f(x)$.

It is convenient to introduce some terminology here. For an interval $m_0 \pm w_0$, we say that the interval is positive if $m_0 - w_0 > 0$, that the interval is negative if $0 > m_0 + w_0$, and that the interval contains 0 if the interval is neither positive nor negative, that is, if $m_0 - w_0 \leq 0 \leq m_0 + w_0$. It is clear then that at any precision of computation, an interval satisfies one of three possibilities: it is positive, it is negative, or it contains 0.

If the $f(x)$ interval $m_0 \pm w_0$ is positive or negative, then it is certain that there is no zero in $[a, b]$. If the $f(x)$ interval contains 0, there may be one or more zeros in $[a, b]$, but we cannot be certain of this, because interval arithmetic does not necessarily yield the true interval but only an interval guaranteed not to be too small. We can get a better idea of where in $[a, b]$ a zero is likely by a simple task queue procedure. Our task queue holds subintervals of $[a, b]$ that must be tested, and initially the task queue holds just the problem interval $[a, b]$. Our first goal is to locate subintervals of $[a, b]$ in which a zero is likely, of wid no larger than some target value W, say W equal to one tenth the $(b - a)/2$ wid of $[a, b]$.

For the first queue subinterval, we compute an interval for $f(x)$ and check whether or not this interval contains 0. If the $f(x)$ interval is positive or negative, the queue subinterval is discarded. If the $f(x)$ interval contains 0, we check whether the queue subinterval's wid is above the target W, and if it is, we bisect the queue subinterval, and the two bisection parts replace it on the queue. If the subinterval's wid is not greater than W, we transfer the subinterval to another list L, initially empty. Eventually the queue becomes empty, and then we can examine the list L. Of course if L is empty, we are done, there being no zeros in $[a, b]$.

Of all the subintervals formed by subdividing $[a, b]$, those containing zeros must appear on the list L. But a subinterval adjoining to one of these may also end up on L, because its computed $f(x)$ interval was wider than the true interval. It is also possible that a zero happens to lie on a boundary point of a subinterval, and then the two subintervals sharing this zero both are on L. To simplify matters, we divide the L subintervals into sets of adjoining subintervals, create for each set a container interval $[c, d]$ equal to the union of all set members, and collect these container intervals in a new list L_1, keeping track of which L members belong to which container interval. A container interval $[c, d]$ on the list L_1 has the property that $f(c)$ is nonzero. If c equals a, the left endpoint of the starting search interval $[a, b]$, this is certainly true. If c is not a, then c, besides being a left endpoint of some subinterval that ended up on the list L, is also a right endpoint of a different subinterval that was discarded. A subinterval is discarded only if its computed $f(x)$ interval is positive or negative, so $f(c)$ is nonzero. Similarly $f(d)$ is nonzero.

Depending on the function $f(x)$ for a container interval $[c, d]$, it may be possible to compute an interval $m_1 \pm w_1$ for $f'(x)$ over $[c, d]$. A favorable circumstance would be finding that the $f'(x)$ interval was positive or negative. In this case, because over $[c, d]$ the function $f(x)$ is either strictly increasing or strictly decreasing, there can be at most one zero in the subinterval, and it is possible to determine if there actually is a zero by comparing the signs of $f(c)$ and $f(d)$. But rather than use this test, which does not generalize easily when we consider finding where several functions of several variables are zero, we use a different test that does generalize easily.

Let $x_0 = (c + d)/2$ be the midpoint of the interval $[c, d]$. According to the Taylor series formula with remainder 4.23, the positive or negative interval $f'(x_0) \pm w_1$ obtained for $f'(x)$ allows the following interval expression for $f(x)$ over $[c, d]$:

$$f(x) = f(x_0) + (f'(x_0) \pm w_1)(x - x_0)$$

If we set $f(x)$ equal to zero and solve for x using interval arithmetic, we obtain the interval

$$x_0 - \frac{f(x_0)}{f'(x_0) \pm w_1} = [c_1, d_1]$$

If this interval lies wholly within $[c, d]$, that is, if we find $c \prec c_1$ and $d_1 \prec d$, then it is certain there is a zero inside $[c, d]$. If the interval $[c_1, d_1]$ lies wholly outside $[c, d]$, that is, if we find $d_1 \prec c$ or we find $d \prec c_1$, then there is no zero inside $[c, d]$.

If a zero lies in $[c, d]$, this zero most often can easily be found to as many places as we please by Newton's method, discussed in the next section. The zero can be identified as a *simple zero*, to indicate a nonzero derivative at the point. If we determine no zero lies in the container interval $[c, d]$, the interval $[c, d]$ and all its associated set members from the list L are discarded.

For some of the container intervals $[c, d]$ on the list L_1, we may find that the interval $[c_1, d_1]$ is neither inside nor outside $[c, d]$, or we may be unable to compute an interval $[c_1, d_1]$ because the $f'(x)$ interval cannot be computed or if computed fails to test positive or negative. For all such cases we move all list L subintervals associated with $[c, d]$ back onto the task queue, and discard $[c, d]$. The target W is reduced by multiplying it by some factor, say $\frac{1}{10}$, and then the entire process of $f(x)$ computation and subinterval bisection already described is repeated, leading to a set of smaller subintervals on L and another list of container intervals on L_1, a repetition of $f'(x)$ computation, and if we are not done, another reduction in the size of W and return of L subintervals to the task queue, and so on. In favorable cases of this problem, all the zeros end up being identified as simple zeros and computed to the required number of places by Newton's method.

The question now is what to do when this favorable outcome does not occur. Each time we examine L_1 there is at least one container interval $[c, d]$ whose $f'(x)$ interval cannot be computed or, if computed, fails to test positive or negative. We may be bounding a zero at which $f'(x)$ is zero or perhaps several closely spaced simple zeros, but it is not even certain that a zero is located. The function $f(x)$ may only come close to zero within the container interval $[c, d]$, never actually becoming zero. For instance, the search interval $[a, b]$ might be $[-1, 1]$ and $f(x)$ might be $x^2 + 10^{-100}$. Here there is no zero, but container intervals surely would be obtained.

It is likely that a container interval $[c, d]$ decreases in length as the parameter W decreases, but again this is not certain. Consider the elementary function

$$f(x) = \min(x, 0) + \max(x - 1, 0) \tag{5.2}$$

which is diagrammed in Fig. 5.1.

This function is zero in the interval $[0, 1]$ and has slope 1 to the right of the interval and to the left. With this function and a search interval $[a, b]$ larger than $[0, 1]$, we would obtain container intervals that did not diminish much in width as the parameter W decreased. However, any point of a container interval at which $f(x) \neq 0$ eventually becomes part of a discarded subinterval as W

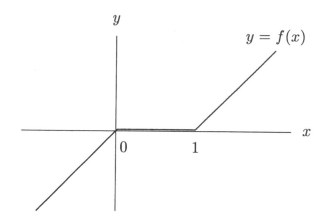

FIGURE 5.1   The function $f(x) = \min(x, 0) + \max(x - 1, 0)$.

decreases, so the points of a container interval have smaller and smaller $|f(x)|$ values. A safe course to take is to continue the cyclic processing of container intervals $[c, d]$ until either of two possibilities occurs:

1. The interval $[c, d]$ becomes so small that it lies within the tilde interval of a k decimal display of its midpoint, and, simultaneously, $|f(x)|$ falls under 10^{-k} for all x in $[c, d]$. In this case the midpoint of this small interval can be displayed to k decimals, the upper bound on $|f(x)|$ given with information about the outcome of the $f'(x)$ computation. If $f(c)$ and $f(d)$ are of opposite signs, it is certain that $[c, d]$ contains at least one zero and that this fact can be communicated, but it is also possible that $[c, d]$ contains an odd number of closely spaced zeros or even a small subinterval $[c_1, d_1]$ on which $f(x)$ is zero. Similarly, if $f(c)$ and $f(d)$ are of the same sign, $[c, d]$ may contain no zero, but there are other possibilities such as $[c, d]$ containing an even number of closely spaced zeros.

2. The interval $[c, d]$ is not small, but it is found that on it $|f(x)|$ is less than some small number, say 10^{-k}. In this case the $|f(x)|$ upper bound of 10^{-k} can be displayed, and the endpoints c and d printed to k decimals to define the interval. If $f(c)$ and $f(d)$ differ in sign, there is definitely at least one zero inside $[c, d]$, and this fact can be communicated.

Now that the various outcomes of a zero search are clearer, we can propose

SOLVABLE PROBLEM 5.1: For any elementary function $f(x)$ defined in an interval $[a, b]$ with $f(a)$ and $f(b)$ nonzero, bound all the zeros of $f(x)$ in $[a, b]$ by giving to k decimals

1. Points identified as simple zeros, or
2. Points containing within their tilde interval a subinterval on which $|f(x)|$ < 10^{-k}, or
3. Endpoints c, d, of subintervals $[c, d]$ on which $|f(x)|$ < 10^{-k}.

For case 2 or 3, a subinterval is identified as *containing at least one zero* if the signs of $f(x)$ differ at the subinterval endpoints.

The $|f(x)|$ magnitude bound of 10^{-k} is arbitrary and can be changed to some other convenient bound, such as 10^{-2k} or 10^{-k^2}.

The results of type 2 or type 3 are escapes from the problem of locating zeros precisely, but for functions $f(x)$, which are allowed to be constant over subintervals, some kind of escape is needed. For instance, suppose we restrict $f(x)$ to be an elementary function for which at least one zero is certain because the signs of $f(a)$ and $f(b)$ differ, and we try to determine just one zero to k places, making no attempt to locate all zeros. If $f(x)$ is unrestricted, even this much simpler problem requires an escape before it becomes solvable. To see this, suppose the search interval is $[-2, 2]$, and our function, diagramed in Fig. 5.2, is

$$f_a(x) = \max(\min(a, x + 1), x - 1)$$

This function has an interval of width 2 on which it equals a and is linear with slope 1 elsewhere. For any value of the parameter a, we have $f(2)$ positive and $f(-2)$ negative. If $a > 0$, there is a single zero at $x = -1$; if $a < 0$, there is a single zero at $x = 1$; and if $a = 0$, all the points in $[-1, 1]$ are zeros. A program that could determine to k decimals a single zero of f_a would allow us

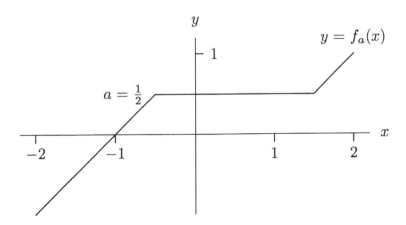

FIGURE 5.2 The function $f_n(x) = \max(\min(a, x + 1), x - 1)$ for $a = \frac{1}{2}$.

to determine from the sign of the supplied zero either that the parameter a satisfied the inequality $a \geq 0$ or that it satisfied $a \leq 0$. However, this contradicts Nonsolvable Problem 3.8, and so we have

NONSOLVABLE PROBLEM 5.2: For any elementary function defined in $[a, b]$, with $f(a)f(b)$ negative, find to k decimals a zero of $f(x)$ in $[a, b]$.

A solvable version of this problem has to allow a response like type 2 or type 3 as an escape.

When solving Problem 5.1 to a particular number of correct decimals, if the information we obtain does not suit our needs, we always can solve the problem to more decimal places. With k larger, we can expect that previously obtained simple zeros get recomputed to more decimals, but with a case 2 or 3 subinterval, there is no telling what might happen. For k larger, the subinterval could yield some simple zeros, disappear altogether, split into several smaller subintervals, or just get smaller.

Now let us reconsider our problem from the standpoint of practicality. It is clear that if the function $f(x)$ is like the function (5.2) and has a subinterval of zero points within $[a, b]$, a dissection of $[a, b]$ into a huge number of small subintervals occurs, and the solution attempt is likely to be computerbound. If we require $f(x)$ to have only isolated points as zeros, then this difficulty does not occur. Another trouble with Problem 5.1 is the requirement that $f(a)$ and $f(b)$ be nonzero. It is likely that we have a particular search interval in mind but have no idea of the value of $f(x)$ at the endpoints of our interval. Why not have the solution program test whether $f(a)$ and $f(b)$ are nonzero (Solvable Problem 3.6 can be used here) and begin a zero search only if these function values test nonzero? With these changes our problem is

SOLVABLE PROBLEM 5.3: For any elementary function $f(x)$ defined in an interval $[a, b]$ having only isolated zeros in $[a, b]$, for a specific positive integer m determine that $|f(a)| < 10^{-m}$ or $|f(b)| < 10^{-m}$ and halt, or, if $f(a)$ and $f(b)$ are nonzero, bound all the zeros of $f(x)$ in $[a, b]$ by giving to k decimals

1. Points identified as simple zeros, or
2. Points containing within their tilde interval a subinterval on which $|f(x)| < 10^{-k}$, the subinterval being identified as *containing at least one zero* if the signs of $f(x)$ differ at the subinterval endpoints.

In our solution procedure, forming a container interval $[c, d]$ requires testing the various subintervals on the list L to find those with matching endpoints. There is no difficulty doing this if the beginning interval $[a, b]$ has exact endpoints a and b, for then we can maintain all derived subinterval endpoints as exact numbers too, and testing these points for equality or inequality poses no difficulty. But what if an endpoint a or b is not exact, perhaps being a number like π or $\cos 15°$? It will be difficult to make correct endpoint comparisons after the subdivision of $[a, b]$ has been in progress a while.

Here we can use an idea employed earlier in Chapter 4 for the problem of determining definite integrals. We map the u interval $[0, 1]$ linearly onto the x interval $[a, b]$, using the relation $x = (1 - u) \cdot a + u \cdot b$, so that we can consider the function f as defined in the u interval $[0, 1]$. A zero located in the u interval is easily converted into a zero in $[a, b]$, and a u subinterval possibly containing a zero is easily converted into a similar x interval. By always using $[0, 1]$ as our search interval, all the subintervals obtained later have exact endpoints too, and the difficulty in comparing endpoints disappears.

5.3. NEWTON'S METHOD

Suppose $f(x)$ is a function with a derivative. Newton's method is an iteration method for progressively improving a zero approximation for such a function. The iteration equation is derived from this simple idea: If we have an approximation x_k to a zero of $f(x)$, then on the graph of $f(x)$ (Fig. 5.3) we can draw a tangent line at the point $x = x_k$ and follow this tangent line to where it crosses the x axis,

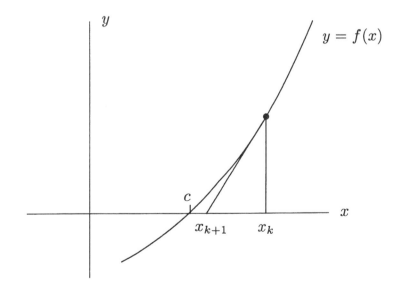

FIGURE 5.3 Newton's method.

obtaining in this way another zero approximation x_{k+1}, hopefully better than x_k. The Taylor series for $f(x)$, with the series expansion point at x_k, is

$$f(x) = f(x_k) + f'(x_k)(x - x_k) + \cdots \tag{5.3}$$

The equation of the tangent line at $x = x_k$ is obtained by discarding all terms of the Taylor series beyond the ones shown to get the straight line equation

$$y = f(x_k) + f'(x_k)(x - x_k)$$

If in this equation we set y equal to zero and solve for x, calling the solution x_{k+1}, then we obtain the equation

$$x_{k+1} = x_k - \frac{f(x_k)}{f'(x_k)} \tag{5.4}$$

The zero finding method described in the preceding section bounds a simple zero in an interval $[c, d]$ where $f'(x)$ is nonzero. Using the iteration Equation 5.4, and with the starting approximation x_0 taken as the midpoint of $[c, d]$, the sequence of iterates generally rapidly locates the zero. However, as Fig. 5.4 shows, it is possible under certain, rare, conditions that the iterates cycle without approaching the zero. If this should occur, we can treat the container interval

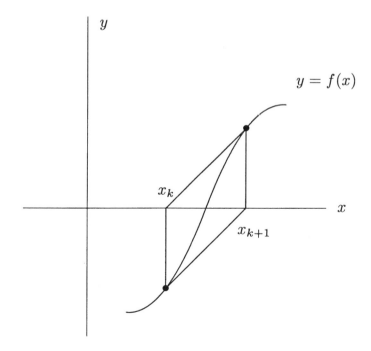

FIGURE 5.4   Cycling with Newton's method.

like a container interval whose $f'(x)$ interval was not of one sign, that is, we discard $[c, d]$ and reprocess its associated subintervals further. Perhaps a smaller container interval leads to success with Newton's method. In the more common case where the iterates approach the zero, we need a criterion for halting the iteration cycle. After an iterate x_{k+1} is computed, it is compared with x_k, and if $x_{k+1} \neq x_k$, then x_{k+1} is replaced by mid x_{k+1} in preparation for the next cycle. Eventually we find $x_{k+1} \doteq x_k$, and then mid x_k is taken as x_z, the final approximation to the zero in $[c, d]$. The case of cycling iterates is detected by finding that $|x_{k+1} - x_k|$ is not decreasing with each iteration.

Next we need to determine the error of x_z, which is the distance from x_z to the zero. For the x_z container interval $[c, d] = x_0 \pm w$, an interval $f'(x_0) \pm w_1$ was computed for $f'(x)$. Suppose we compute the interval $y = f(x_z)/(f'(x_0) \pm w_1)$. Then, as we show, the error of x_z is bounded by $e = |\text{mid } y| + \text{wid } y$, so $x_z \oplus e$ may be taken as a correctly ranged zero approximation. This result follows from the mean value theorem. At the true zero z, we have $f(z) = 0$, so $f(x_z) = f(x_z) - f(z) = f'(c_x)(x_z - z)$, where the point c_x is between x_z and z. The error $|x_z - z|$ equals $|f(x_z)|/|f'(c_x)|$, and because $f'(c_x)$ lies in the interval $f'(x_0) \pm w_1$, the error is bounded by e. If $x_z \oplus e$ does not give us the desired number of correct decimal places we want, we need to increase the precision of computation, restart the Newton's method iteration process, and repeat the error test with a better x_z.

5.4. ORDER OF CONVERGENCE

In the later cycles of Newton's method, when x_k gets close to a zero z the convergence accelerates, with the number of correct decimal places of the iterates approximately doubling at each step. We can show this by using the Taylor series remainder formula, adding to the terms shown in equation (5.3) an additional remainder term:

$$f(x) = f(x_k) + f'(x_k)(x - x_k) + \frac{1}{2}f''(c_x)(x - x_k)^2$$

Here c_x is some point between x and x_k. If we set x equal to the zero z, then we have

$$0 = f(x_k) + f'(x_k)(z - x_k) + \frac{1}{2}f''(c_z)(z - x_k)^2$$

When we divide this equation by $f'(x_k)$ and make use of the Newton iteration equation, we find

$$z - x_{k+1} = -\frac{f''(c_z)}{2f'(x_k)}(z - x_k)^2 \tag{5.5}$$

Because the term in $z - x_k$ is squared, if $f'(z) \neq 0$ and x_k is close enough to z, then x_{k+1} is even closer to z, which implies that the iterates after x_k converge to z.

The quantity $|z - x_k|$ is the error of the approximation x_k. Taking absolute values of both sides of Equation 5.5, we obtain

$$\text{Error } x_{k+1} = L_k (\text{Error } x_k)^2$$

where $L_k = |f''(c_z)/2f'(x_k)|$. The coefficient L_k approaches a limiting value L because

$$\lim_{k \to \infty} L_k = \lim_{x_k \to z} \frac{|f''(c_z)|}{2|f'(x_k)|} = \frac{|f''(z)|}{2|f'(z)|} = L$$

Thus in the last stages of Newton's method, the equation below holds very nearly.

$$\text{Error } x_{k+1} = L(\text{Error } x_k)^2$$

From this equation, the number of correct decimal places approximately doubles at each step and may be more than this or less depending on the magnitude of L.

When a series of approximates x_k approaches the true value z, such that the error satisfies the relation

$$\lim_{k \to \infty} \frac{\text{Error } x_{k+1}}{(\text{Error } x_k)^s} = L > 0$$

the convergence is said to be *of order s*. Thus the convergence of the iterates x_k of Newton's method is generally of order two, or x_k converges *quadratically*. If the exponent s equals one, then x_k is said to converge *linearly*. For linear convergence, the positive constant L in the limit above must be less than 1 and the nearer L is to 1, the slower the convergence.

Linear convergence is encountered with Newton's method if the derivative at z is zero. Suppose we have $f'(z) = f''(z) = \ldots = f^{(k-1)}(z) = 0$, and $f^{(k)}(z) \neq 0$. If we expand $f(x)$ in a Taylor series about the zero point z and use the Taylor series remainder formula, at $x = x_k$ we have the equation

$$f(x_k) = f(z) + f'(z)(x_k - z) + \frac{1}{2!}f''(z)(x_k - z)^2 + \cdots + \frac{1}{k!}f^{(k)}(c_1)(x_k - z)^k$$

$$= \frac{1}{k!}f^{(k)}(c_1)(x_k - z)^k$$

Here c_1 is some point between x_k and z. Similarly, for $f'(x_k)$ we have the equation

$$f'(x_k) = \frac{1}{(k-1)!}f^{(k)}(c_2)(x_k - z)^{k-1}$$

where c_2 is some point between x_k and z. We have then

$$x_{k+1} - z = x_k - \frac{f(x_k)}{f'(x_k)} - z = (x_k - z) - \frac{f^{(k)}(c_1)}{kf^{(k)}(c_2)}(x_k - z)$$

If x_k is close enough to z, then $\frac{f^{(k)}(c_1)}{kf^{(k)}(c_2)}$ is near to $\frac{1}{k}$, and convergence becomes certain because $|x_{k+1} - z|$ is less than $|x_k - z|$. Then x_k approaches z in the limit, and we have

$$\lim_{k \to \infty} \frac{\text{Error } x_{k+1}}{\text{Error } x_k} = \lim_{x_k \to z} \left| 1 - \frac{f^{(k)}(c_1)}{kf^{(k)}(c_2)} \right| = \left| 1 - \frac{f^{(k)}(z)}{kf^{(k)}(z)} \right| = \frac{k-1}{k}$$

The convergence is linear and is slower for larger k.

5.5. EXTENDING THE SOLUTION METHOD TO THE GENERAL PROBLEM

For the general problem we employ vector notation. We use **x** to denote a vector with components $x_1, x_2, \ldots, x_n$. Such a vector is determined *to k decimals* by giving all its components to k decimals. The number of components of our vectors is understood to always equal n, which is the number of functions f_i and the number of variables x_i. We have now a vector-valued function **f(x)** with components

$$f_i(\mathbf{x}) = f_i(x_1, x_2, \ldots, x_n), \qquad i = 1, 2, \ldots, n$$

The function **f(x)** is called elementary if all component functions are elementary. The search region is defined by the relations $\mathbf{a} \leq \mathbf{x} \leq \mathbf{b}$, where the constant vector **a** has components equal to the left endpoints of the n intervals defining the region, and **b** has components equal to the right endpoints. Here the relation $\mathbf{x} \leq \mathbf{y}$ denotes the n component relations $x_i \leq y_i$. When n equals three, the search region is a *box* in the argument space. In general, for any n, we use the convenient term *box* for the search region. A box B is located to k decimals by giving its vectors **a** and **b** to k decimals, and the wid of the box is $\max_j \frac{b_j - a_j}{2}$. The zero problem now is that of finding in a specified box B the arguments **x** where **f(x)** = **0**, with **0** of course denoting a vector with all components zero.

When n is 1, a connected set of zeros must be an interval, that is, a box of dimension 1. When n is greater than 1, it is possible to have a connected set of zeros that is not a box. For instance, if $f_1(x_1, x_2) = f_2(x_1, x_2) = x_1^2 + x_2^2 - 1$, then all points on the unit circle are zeros.

The method described in Section 5.2 for the case $n = 1$ can be generalized to find zeros for any n. However, there are some items of that single function procedure that need to be reinterpreted. If the $f'(x)$ interval is of one sign in a subinterval $[c, d]$, this indicates that there can be at most one zero in $[c, d]$. To obtain a similar inference for $\mathbf{f}(\mathbf{x})$ over a subbox $B^{(S)}$, we must compute an interval for the Jacobian over $B^{(S)}$. The Jacobian at a point $\mathbf{x}$ is the determinant of the matrix $J(\mathbf{x})$ defined by the equation

$$
J(\mathbf{x}) = \begin{bmatrix}
\dfrac{\partial f_1(\mathbf{x})}{\partial x_1} & \dfrac{\partial f_1(\mathbf{x})}{\partial x_2} & \cdots & \dfrac{\partial f_1(\mathbf{x})}{\partial x_n} \\
\dfrac{\partial f_2(\mathbf{x})}{\partial x_1} & \dfrac{\partial f_2(\mathbf{x})}{\partial x_2} & \cdots & \dfrac{\partial f_2(\mathbf{x})}{\partial x_n} \\
\vdots & \vdots & \vdots\vdots & \vdots \\
\dfrac{\partial f_n(\mathbf{x})}{\partial x_1} & \dfrac{\partial f_n(\mathbf{x})}{\partial x_2} & \cdots & \dfrac{\partial f_n(\mathbf{x})}{\partial x_n}
\end{bmatrix}
$$

(When n is 1, the Jacobian is the derivative $\frac{df_1(x_1)}{dx_1}$.) If $\mathbf{f}(\mathbf{x})$ is such that we can compute intervals over $B^{(S)}$ for all the $J(\mathbf{x})$ elements, then we also can compute an interval for det $J(\mathbf{x})$, the Jacobian. If we find that this interval is positive or negative, then there can be at most one zero in $B^{(S)}$. To show this we need the following well-known result.

THEOREM 5.1: *Let the box B be defined by $\mathbf{a} \leq \mathbf{x} \leq \mathbf{b}$ where $\mathbf{a}$ and $\mathbf{b}$ are two constant vectors. If $\mathbf{f}(\mathbf{x})$ is defined in B and has all first partial derivatives there, then if $\mathbf{y}$ and $\mathbf{z}$ are any two points in B, there are points $\mathbf{c}_1, \mathbf{c}_2, \ldots, \mathbf{c}_n$ along the line segment joining $\mathbf{y}$ and $\mathbf{z}$ such that*

$$
f_i(\mathbf{z}) - f_i(\mathbf{y}) = \sum_{j=1}^{n} \frac{\partial f_i(\mathbf{c}_i)}{\partial x_j}(z_j - y_j), \qquad i = 1, \ldots, n
$$

For each i we may define a function $F_i(t) = f_i((1 - t)\mathbf{y} + t\mathbf{z})$ with the argument t varying in $[0, 1]$. Making use of the mean value theorem, we have

$$
F_i(1) - F_i(0) = f_i(\mathbf{z}) - f_i(\mathbf{y}) = \frac{dF_i(t_c)}{dt}(1 - 0) = \sum_{j=1}^{n} \frac{\partial f_i((1 - t_c)\mathbf{y} + t_c\mathbf{z})}{\partial x_j}(z_j - y_j)
$$

$$
= \sum_{j=1}^{n} \frac{\partial f_i(\mathbf{c}_i)}{\partial x_j}(z_j - y_j)
$$

where $\mathbf{c}_i$ equals $(1 - t_c)\mathbf{y} + t_c\mathbf{z}$, and t_c is a point in $(0, 1)$, possibly different for each i. This proves the theorem.

COROLLARY: *If the computed Jacobian interval over B is of one sign, that is, does not contain 0, then $\mathbf{f}(\mathbf{x})$ is one-to-one over B, and as a consequence, there can be at most one zero in B.*

If $\mathbf{y}$ and $\mathbf{z}$ were two distinct arguments in B with $\mathbf{f}(\mathbf{y}) = \mathbf{f}(\mathbf{z})$, it would follow from the theorem that there were points $\mathbf{c}_1, \mathbf{c}_2, \ldots, \mathbf{c}_n$ inside B such that

$$
\begin{bmatrix}
\dfrac{\partial f_1(\mathbf{c}_1)}{\partial x_1} & \dfrac{\partial f_1(\mathbf{c}_1)}{\partial x_2} & \cdots & \dfrac{\partial f_1(\mathbf{c}_1)}{\partial x_n} \\[2mm]
\dfrac{\partial f_2(\mathbf{c}_2)}{\partial x_1} & \dfrac{\partial f_2(\mathbf{c}_2)}{\partial x_2} & \cdots & \dfrac{\partial f_2(\mathbf{c}_2)}{\partial x_n} \\[2mm]
\vdots & \vdots & \vdots\vdots & \vdots \\[2mm]
\dfrac{\partial f_n(\mathbf{c}_n)}{\partial x_1} & \dfrac{\partial f_n(\mathbf{c}_n)}{\partial x_2} & \cdots & \dfrac{\partial f_n(\mathbf{c}_n)}{\partial x_n}
\end{bmatrix}
\begin{bmatrix}
z_1 - y_1 \\
z_2 - y_2 \\
\vdots \\
z_n - y_n
\end{bmatrix}
=
\begin{bmatrix}
0 \\
0 \\
\vdots \\
0
\end{bmatrix}
$$

The determinant of the matrix above must be zero, because the product of the matrix with the nonzero vector $\mathbf{z} - \mathbf{y}$ is $\mathbf{0}$. On the other hand, this determinant is a Jacobian evaluated at a point inside B, so each matrix element is inside the corresponding partial derivative interval of the matrix $J(\mathbf{x})$ over B, which contradicts the assumption that the Jacobian interval does not contain 0.

In the zero finding procedure for $f(x)$, it is possible to determine whether a container interval $[c, d]$ has at least one $f(x)$ zero, by examining the signs of $f(x)$ at the boundary points of $[c, d]$. This procedure must be generalized. We need to be able to determine whether a box B contains at least one zero of $\mathbf{f}(\mathbf{x})$ by examining $\mathbf{f}(\mathbf{x})$ on the boundary of B. If $\mathbf{f}(\mathbf{x}) \neq \mathbf{0}$ on the boundary of B, there is a certain integer that can be computed, called the *topological degree*. If the topological degree is unequal to zero (analogous to the $f(x)$ case where the signs of $f(a)$ and $f(b)$ differ), there is at least one zero inside B. In the next two sections we describe the topological degree in more detail and give a method for computing it.

Now we can generalize the $f(x)$ solvable problems:

SOLVABLE PROBLEM 5.4: For any elementary function $\mathbf{f}(\mathbf{x})$ defined on a box B with $\mathbf{f}(\mathbf{x}) \neq \mathbf{0}$ on the boundary of B, bound all the zeros of $\mathbf{f}$ in B by giving to k decimals

1. Points identified as simple zeros, or
2. Points containing within their tilde box a region where $\max_i |f_i(\mathbf{x})| < 10^{-k}$, or
3. Subboxes each bounding a region where $\max_i |f_i(\mathbf{x})| < 10^{-k}$.

The regions in 2 and 3 are identified as *containing at least one zero* if the $\mathbf{f}(\mathbf{x})$ topological degree over a bounding box is nonzero. (The tilde box of a point $\mathbf{x}$ is the box determined by the tilde intervals of the components x_i.)

We have a more practical version of our problem if we make changes similar to those made to Problem 5.1 to obtain Problem 5.3.

SOLVABLE PROBLEM 5.5: For any elementary function $\mathbf{f}(\mathbf{x})$ defined on a box B, having only isolated zeros in B, for a specific positive integer m, locate to m decimals a point on the boundary of B where all components of $\mathbf{f}(\mathbf{x})$ are less in magnitude than 10^{-m} and halt, or, if $\mathbf{f}(\mathbf{x}) \neq \mathbf{0}$ on the boundary of B, bound all the zeros of $\mathbf{f}$ in B by giving to k decimals

1. Points identified as simple zeros, or
2. Points containing within their tilde box a region where $\max_i |f_i(\mathbf{x})| < 10^{-k}$, the region identified as *containing at least one zero* if the $\mathbf{f}(\mathbf{x})$ topological degree over a bounding box is nonzero.

We describe now the general computation scheme for solving this problem. In the computation it is helpful to use a reference box $B^{(\mathbf{u})}$, with all dimension intervals being $[0, 1]$, the i-th interval with variable u_i being mapped linearly into the corresponding B interval $[a_i, b_i]$ with variable x_i, according to the rule

$$x_i = (1 - u_i)a_i + u_i b_i$$

Any required subdivision of B is done by making the appropriate subdivision of the reference box $B^{(\mathbf{u})}$, with each interval endpoint of a $B^{(\mathbf{u})}$ subbox maintained as an exact ranged number. The interval endpoints for the corresponding B subbox are obtained by using the linear relations above. Controlling the subdivision of B this way has two advantages: If the precision of computation is increased, the endpoints of the intervals defining a B subbox are obtained to higher precision too. It is always possible to determine when two subboxes of B are adjoining by comparing the exact interval endpoints of their $B^{(\mathbf{u})}$ reference subboxes. In our description of the zero finding procedure, each subbox of B is to be understood as defined by a corresponding subbox of $B^{(\mathbf{u})}$.

The first step of the general procedure is to test $\mathbf{f}(\mathbf{x})$ to make certain it is defined over B. We need here a task queue holding boxes to be checked. Initially the queue holds just the starting box B. The task queue cycle with the first queue box $B^{(Q)}$ is as follows: We set all series primitives x_j to their box interval values and then generate intervals for all component functions f_i. If there are no series errors, the subbox is discarded. If there is an error in the evaluation of a component, we construct all possible subboxes that can be formed by bisecting all $B^{(Q)}$ dimension intervals, and these 2^n subboxes replace $B^{(Q)}$ on the queue. Eventually, either the queue becomes empty, in which case $\mathbf{f}(\mathbf{x})$ passes its test, or else the leading queue box becomes small enough to fit inside the tilde-box of its centerpoint, if it were displayed to m decimals. The box centerpoint can now be displayed to m decimals to indicate a point in the proposed search box B where some component of $\mathbf{f}$ can not be computed. As was indicated in Section 3.6, no function test can be completely foolproof. It is always possible to get certain improperly specified functions past defenses. The consequences are *undefined*, and the failure of the test is not construed as a failure in treating the solvable problem.

Next we test **f** over the boundary of B to make certain that the components of **f(x)** are nonzero. Again a task queue is needed, but for this examination we must allow the queue to hold domains that define a side of a box or a part of a side. A queue domain now has n fields s_i, $i = 1, \ldots, n$, each field being flagged either as a *point* holding a single number a, or as an *interval* holding a number pair a, b, with $a < b$. We distinguish the two cases by writing $s_i = a$ or $s_i = [a, b]$. The queue initially has $2n$ domains defining the boundary of B. For instance, the two domains specifying the sides of B with x_j fixed are obtained by setting

$$s_i = \begin{cases} a_j \text{ or } b_j & \text{for } i = j \\ [a_i, b_i] & \text{for } i \neq j \end{cases}$$

The task queue cycle with the leading queue domain is as follows: We use the s_i fields to set the x_i primitives appropriately, a *point* field resulting in a zero wid, and then obtain intervals for all components f_k. If any of these intervals does not contain 0, the domain is discarded. On the other hand, if for all components f_k we either get an interval containing 0 or get a series evaluation error, then we construct all possible subdomains that can be formed by bisecting *interval* dimensions, the point dimension merely being copied, and these 2^{n-1} subdomains replace the leading queue domain. Eventually, either the queue becomes empty, in which case the test is passed, or the leading queue domain fails within the *tilde* box of its centerpoint if it were displayed to m decimals. When this happens, the centerpoint is displayed to m decimals as a point on the boundary of B where either some component of **f(x)** is undefined or where the test **f(x)** $\neq$ **0** fails.

Once these two tests are passed, the next phase is to get a list of B subboxes where a zero is likely. The target wid W is set to an initial value, say one tenth the B wid. Our first goal will be to find B subboxes where a zero is likely, the wid of the subbox being no larger than W. Again we need a task queue holding subboxes, and initially the queue holds just the box B. The task queue cycle is as follows: We set the primitives x_j to the component intervals of the first queue box $B^{(Q)}$ and then attempt to generate intervals for the functions f_i. If we obtain a series evaluation error, then $B^{(Q)}$ is divided into two subboxes by bisecting its largest component interval, and these two subboxes replace $B^{(Q)}$ on the queue. If we obtain intervals for all functions f_i, and one of these intervals does not contain 0, then $B^{(Q)}$ is discarded. If all f_i intervals contain 0, the wid of $B^{(Q)}$ is compared with the control parameter W, and if it is larger, then $B^{(Q)}$ is divided into two subboxes by bisecting its largest component interval, and these two subboxes replace $B^{(Q)}$ on the queue. Otherwise $B^{(Q)}$ is moved to another list L, initially empty, for later processing. Eventually the queue becomes empty. If the list L is also empty we are done, there being no zeros in B.

The next part of the procedure is to arrange the list of subboxes on L into sets of boxes that are mutually adjoining. (Two boxes of dimension n adjoin if the intersection of their boundaries is a box of dimension $n - 1$.) Any subbox belonging to one of these sets is connected to any other subbox of the set either directly or via other adjoining subboxes of the set. For each set we make up a container box $B^{(C)}$ just large enough, in each dimension, to contain all of the subboxes of a set. With this collection of container boxes, it is certain that $\mathbf{f}$ is not equal to $\mathbf{0}$ on the boundary of any box $B^{(C)}$. Any point $\mathbf{x}$ on the boundary of $B^{(C)}$ is either a boundary point of B, implying $\mathbf{f}(\mathbf{x}) \neq \mathbf{0}$, or is also a boundary point of some adjoining subbox that got discarded in the testing process, again implying $\mathbf{f}(\mathbf{x}) \neq \mathbf{0}$.

Next a Jacobian test is made for each container box $B^{(C)}$. Let $\mathbf{x}^{(C)}$ be the centerpoint of $B^{(C)}$. Here we attempt to obtain intervals for both $f_i(\mathbf{x})$ and the partial derivatives $\partial f_i(\mathbf{x})/\partial x_j$ over $B^{(C)}$. If these are obtained, then using the remainder form for multiple variable series expansions given in Section 4.9, we can write the interval relations

$$f_1(\mathbf{x}) = f_1(\mathbf{x}^{(C)}) + \sum_{j=1}^{n} \left(\frac{\partial f_1(\mathbf{x}^{(C)})}{\partial x_j} \pm w_{1j} \right)(x_j - x_j^{(C)})$$

$$f_2(\mathbf{x}) = f_2(\mathbf{x}^{(C)}) + \sum_{j=1}^{n} \left(\frac{\partial f_2(\mathbf{x}^{(C)})}{\partial x_j} \pm w_{2j} \right)(x_j - x_j^{(C)})$$

$$\vdots \qquad \vdots$$

$$f_n(\mathbf{x}) = f_n(\mathbf{x}^{(C)}) + \sum_{j=1}^{n} \left(\frac{\partial f_n(\mathbf{x}^{(C)})}{\partial x_j} \pm w_{nj} \right)(x_j - x_j^{(C)})$$

These equations can be written in vector–matrix notation as

$$\mathbf{f}(\mathbf{x}) = \mathbf{f}(\mathbf{x}^{(C)}) + [J(\mathbf{x}^{(C)}) \pm W](\mathbf{x} - \mathbf{x}^{(C)}) \tag{5.6}$$

We set $\mathbf{f}(\mathbf{x})$ equal to $\mathbf{0}$, and attempt to solve the system 5.6 for $\mathbf{x}$, using an interval version of Gaussian elimination. If the process is successful, the Jacobian interval definitely is positive or negative, and we can determine if a zero lies inside $B^{(C)}$ by an examination of the $\mathbf{x}$ solution, which consists of n intervals defining some box $\hat{B}$. If this box is contained in $B^{(C)}$, there is a zero in $B^{(C)}$, and it can usually be found easily to any number of decimal places by the generalized Newton's method, discussed in Section 5.8. If the box $\hat{B}$ is outside $B^{(C)}$, it is certain there is no zero in $B^{(C)}$, and the box $B^{(C)}$ and all its associated subboxes are discarded.

If neither of these two cases occurs, whether there is a zero in $B^{(C)}$ is undecided. For any such box, and also for a container box for which the described procedure cannot be carried through for some reason, the box size is checked to determine whether it has shrunk to the size required by outcome 2 of Problem 5.5. If it has, then we test whether for all i the inequality $|f_i(\mathbf{x})| < 10^{-k}$ is true on each list L subbox associated with $B^{(C)}$. If these tests are passed, the topological degree is computed for $\mathbf{f}(\mathbf{x})$ over $B^{(C)}$, and then the container box $B^{(C)}$ and its associated subboxes are discarded after an appropriate display of information. Any container boxes failing these tests are discarded, after first moving their associated subboxes back to the task queue. The parameter W is reduced by some factor, say $\frac{1}{10}$, and another cycle of subbox processing ensues.

5.6. THE TOPOLOGICAL DEGREE

We define the topological degree for elementary functions as a certain surface integral, first introduced by Kronecker. Actually the integral is over an $(n - 1)$-dimensional set in n dimensional space, but, for the sake of simplicity, we always use the terms *surface area* and *surface integral* and other expressions that apply only to the three-dimensional situation. We review how to compute surface areas and surface integrals in n dimensional space.

A *smooth* surface is defined by a set of n functions $g_1, \ldots, g_n$ of $n - 1$ variables $u_1, \ldots, u_{n-1}$, defined over some box of $(n - 1)$-dimensional Euclidean space, with all the functions having partial derivatives with respect to the u_i variables. We will suppose that all our functions g_i are elementary. We recall that a signed n-dimensional volume of the n-dimensional parallelepiped formed by n linearly independent vectors $\mathbf{a}^{(1)}, \mathbf{a}^{(2)}, \ldots, \mathbf{a}^{(n)}$ is given by $\det(\mathbf{a}^{(1)}, \mathbf{a}^{(2)}, \ldots, \mathbf{a}^{(n)})$ where

$$\det(\mathbf{a}^{(1)}, \mathbf{a}^{(2)}, \ldots, \mathbf{a}^{(n)}) = \det \begin{bmatrix} a_1^{(1)} & a_2^{(1)} & \cdots & a_n^{(1)} \\ a_1^{(2)} & a_2^{(2)} & \cdots & a_n^{(2)} \\ \vdots & \vdots & \vdots\vdots\vdots & \vdots \\ a_1^{(n)} & a_2^{(n)} & \cdots & a_n^{(n)} \end{bmatrix}$$

If the sign of $\det(\mathbf{a}^{(1)}, \mathbf{a}^{(2)}, \ldots, \mathbf{a}^{(n)})$ is $+$, the vectors $\mathbf{a}^{(1)}, \mathbf{a}^{(2)}, \ldots, \mathbf{a}^{(n)}$ are said to be right-handed; if the sign is $-$, they are left-handed. Of course if any two vectors $\mathbf{a}^{(i)}$ and $\mathbf{a}^{(j)}$ in the vector list are exchanged, the handedness changes.

Now let $\mathbf{e}_1, \mathbf{e}_2, \ldots, \mathbf{e}_n$ be unit basis vectors associated with our n dimensional coordinate system $x_1, x_2, \ldots, x_n$. That is, $\mathbf{x} = \Sigma_i x_i \mathbf{e}_i$. Consider the vector

$$
\mathbf{Y_g} = \det
\begin{bmatrix}
\mathbf{e}_1 & \mathbf{e}_2 & \cdots & \mathbf{e}_n \\
\dfrac{\partial g_1}{\partial u_1} & \dfrac{\partial g_2}{\partial u_1} & \cdots & \dfrac{\partial g_n}{\partial u_1} \\
\dfrac{\partial g_1}{\partial u_2} & \dfrac{\partial g_2}{\partial u_2} & \cdots & \dfrac{\partial g_n}{\partial u_2} \\
\vdots & \vdots & \vdots\vdots\vdots & \vdots \\
\dfrac{\partial g_1}{\partial u_{n-1}} & \dfrac{\partial g_2}{\partial u_{n-1}} & \cdots & \dfrac{\partial g_n}{\partial u_{n-1}}
\end{bmatrix}
= \sum_{i=1}^{n} y_i \mathbf{e}_i
\tag{5.7}
$$

The vector $\mathbf{Y_g}$ is orthogonal to the vectors $\dfrac{\partial \mathbf{g}}{\partial u_1}, \dfrac{\partial \mathbf{g}}{\partial u_2}, \ldots, \dfrac{\partial \mathbf{g}}{\partial u_{n-1}}$ defined by rows 2 through n of the matrix above, because the dot product of any of these with $\mathbf{Y_g}$ results in an ordinary (nonvectorial) determinant with two identical rows. The $(n-1)$-dimensional area of the $(n-1)$-dimensional parallelogram formed by $\dfrac{\partial \mathbf{g}}{\partial u_1}, \ldots, \dfrac{\partial \mathbf{g}}{\partial u_{n-1}}$ can be obtained as the volume $\det(\dfrac{1}{|\mathbf{Y_g}|} \mathbf{Y_g}, \dfrac{\partial \mathbf{g}}{\partial u_1}, \ldots, \dfrac{\partial \mathbf{g}}{\partial u_{n-1}})$, which equals

$$
\det
\begin{bmatrix}
\dfrac{y_1}{|\mathbf{Y_g}|} & \dfrac{y_2}{|\mathbf{Y_g}|} & \cdots & \dfrac{y_n}{|\mathbf{Y_g}|} \\
\dfrac{\partial g_1}{\partial u_1} & \dfrac{\partial g_2}{\partial u_1} & \cdots & \dfrac{\partial g_n}{\partial u_1} \\
\dfrac{\partial g_1}{\partial u_2} & \dfrac{\partial g_2}{\partial u_2} & \cdots & \dfrac{\partial g_n}{\partial u_2} \\
\vdots & \vdots & \vdots\vdots\vdots & \vdots \\
\dfrac{\partial g_1}{\partial u_{n-1}} & \dfrac{\partial g_2}{\partial u_{n-1}} & \cdots & \dfrac{\partial g_n}{\partial u_{n-1}}
\end{bmatrix}
= \dfrac{1}{|\mathbf{Y_g}|} \sum_{i-1}^{n} y_i^2 = |\mathbf{Y_g}|
$$

Thus the vectors $\mathbf{Y_g}, \dfrac{\partial \mathbf{g}}{\partial u_1}, \dfrac{\partial \mathbf{g}}{\partial u_2}, \ldots, \dfrac{\partial \mathbf{g}}{\partial u_{n-1}}$ are right-handed, and the length of $\mathbf{Y_g}$ equals the $(n-1)$-dimensional area of the $\dfrac{\partial \mathbf{g}}{\partial u_1}, \dfrac{\partial \mathbf{g}}{\partial u_2}, \ldots, \dfrac{\partial \mathbf{g}}{\partial u_{n-1}}$ parallelogram. We see then that the surface area of $\mathbf{g}$ is

$$
\int |\mathbf{Y_g}| \, du_1 \cdots du_{n-1}
\tag{5.8}
$$

with the domain of integration being the box in which $\mathbf{g}$ is defined. If there is a vector field $\mathbf{V(x)}$ defined over the surface, it is possible to define a *flux* integral of $\mathbf{V}$ over the surface as

$$
\int \mathbf{V} \cdot \mathbf{Y_g} du_1 \cdots du_{n-1}
$$

In a calculus course, the surface area integral 5.8 is usually encountered for the cases $n = 2$ or $n = 3$. When n is 2, the surface area (actually arc length) equals

$$\int \sqrt{\left(\frac{dg_1}{du_1}\right)^2 + \left(\frac{dg_2}{du_1}\right)^2} \, du_1$$

When n is 3, the surface area equals

$$\int \left|\frac{\partial \mathbf{g}}{\partial u_1} \times \frac{\partial \mathbf{g}}{\partial u_2}\right| du_1 du_2$$

$$= \int \sqrt{\left[\frac{\partial(g_1, g_2)}{\partial(u_1, u_2)}\right]^2 + \left[\frac{\partial(g_2, g_3)}{\partial(u_1, u_2)}\right]^2 + \left[\frac{\partial(g_3, g_1)}{\partial(u_1, u_2)}\right]^2} \, du_1 du_2$$

Here we are encountering for the first time a widely employed Jacobian notation. The symbol $\frac{\partial(g_1, g_2, \ldots, g_n)}{\partial(u_1, u_2, \ldots, u_n)}$ is defined by the equation

$$\frac{\partial(g_1, g_2, \ldots, g_n)}{\partial(u_1, u_2, \ldots, u_n)} = \det \begin{bmatrix} \frac{\partial g_1}{\partial u_1} & \frac{\partial g_1}{\partial u_2} & \cdots & \frac{\partial g_1}{\partial u_n} \\ \frac{\partial g_2}{\partial u_1} & \frac{\partial g_2}{\partial u_2} & \cdots & \frac{\partial g_2}{\partial u_n} \\ \vdots & \vdots & \vdots\vdots\vdots & \vdots \\ \frac{\partial g_n}{\partial u_1} & \frac{\partial g_n}{\partial u_2} & \cdots & \frac{\partial g_n}{\partial u_n} \end{bmatrix}$$

Now suppose we have a connected closed region D of n-space whose boundary, ∂D, is *piecewise smooth* and elementary, that is, ∂D can be defined as a single smooth elementary surface or as a finite number of smooth elementary surfaces. For instance, if D is a sphere, only one elementary surface is needed to define ∂D, while if D is a box, six surfaces are needed to define the six sides of D. Suppose we arrange the order of the variables $u_1, \ldots, u_{n-1}$ defining the smooth surfaces so that the vectors $\mathbf{Y_g}$ always point outward from D. We can always arrange this for $n > 2$ by renumbering, if necessary, the u_i variables. If $n = 2$, there is only a single variable u_1, and here we may need to replace u_1 by $\hat{u}_1 = -u_1$ to get $\mathbf{Y_g}$ to point outward from D.

If the origin, that is, the point $\mathbf{0}$, does not lie on ∂D, it is possible to define a flux integral over ∂D that can equal only 1 or 0, according as D does or does not contain $\mathbf{0}$. The general idea is to project the surface ∂D onto the unit n dimensional sphere S_n, defined by the equation $\sum_{i=1}^n x_i^2 = 1$, using rays from the origin to effect the projection, and to compare the area of S_n covered by this projection with the total area of S_n. Let P be a point on ∂D with coordinates

$g_i(u_1, \ldots, u_{n-1})$ determining the vector $\mathbf{g}$ stretching from the origin to P. The surface element $|\mathbf{Y_g}|du_1 \ldots du_{n-1}$ is changed before projection to $\frac{1}{|\mathbf{g}|}\mathbf{g} \cdot \mathbf{Y_g}\, du_1 \ldots du_{n-1}$ to reflect the direction of $\mathbf{g}$, which is perpendicular to S_n. Then this result is multiplied by the factor $\frac{1}{|\mathbf{g}|^{n-1}}$ for scaling the $(n-1)$-dimensional area properly for projection to S_n. Let $A(S_n)$ be the surface area of S_n. For instance, we have $A(S_2) = 2\pi$, and $A(S_3) = 4\pi$. The flux surface integral we want is

$$\frac{1}{A(S_n)} \int \frac{1}{|\mathbf{g}|^n}\, \mathbf{g} \cdot \mathbf{Y_g}\, du_1 \cdots u_{n-1}$$

$$= \frac{1}{A(S_n)} \int \frac{1}{|\mathbf{g}|^n} \det\left(\mathbf{g}, \frac{\partial \mathbf{g}}{\partial u_1}, \ldots, \frac{\partial \mathbf{g}}{\partial u_{n-1}}\right) du_1 \cdots du_{n-1} \tag{5.9}$$

Note that a projected surface element from a point $\mathbf{g}$ on ∂D is counted positive or negative according as the component of $\mathbf{Y_g}$ in the direction of $\mathbf{g}$ is positive or negative, respectively.

Suppose now that $\mathbf{0}$ lies inside D. Every ray from $\mathbf{0}$ pierces the surface ∂D in one or more points. Let N_+ be the number of piercings with a positive projected surface element, and N_- be the number with a negative projected surface element. We can ignore any points where the ray is tangent to ∂D because then the projected surface element has a multiplier of zero. A ray cannot have an infinite number of isolated piercing points with an *elementary* surface. No matter how complicated the surface ∂D is, the difference $N_+ - N_-$ always is 1 for a ray passing through any point of S_n, and the integral 5.9 equals one. On the other hand, if $\mathbf{0}$ is outside D, the difference $N_+ - N_-$ is always 0, and the integral equals zero. Here the arrangement of having $\mathbf{Y_g}$ point outwards from D is crucial.

Now consider an elementary function $\mathbf{f}$ defined over the region D. The function $\mathbf{f}$ maps the surface ∂D onto another surface $\mathbf{f}(\partial D)$. The surface $\mathbf{f}(\partial D)$ is defined by the composition of $\mathbf{f}$ with the various functions $\mathbf{x} = \mathbf{g}(u_1, \ldots, u_{n-1})$ defining ∂D. This surface may intersect itself. We take as our definition of the topological degree, the preceding integral over this surface. That is, following Kronecker, we have

$$td(\mathbf{f}, D) = \frac{1}{A(S_n)} \int \frac{1}{|\mathbf{f(g)}|^n} \det\left(\mathbf{f(g)}, \frac{\partial \mathbf{f(g)}}{\partial u_1} \cdots \frac{\partial \mathbf{f(g)}}{\partial u_{n-1}}\right) du_1 \cdots du_{n-1} \tag{5.10}$$

In order that this integral be defined, no point of ∂D can be mapped by $\mathbf{f}$ into $\mathbf{0}$. It is also necessary that all the partial derivatives of $\mathbf{f}$ be defined, except perhaps on a set small enough so that the integral still can be obtained as an appropriate *improper integral*.

Because the vector $\mathbf{Y_g}$ points outward from D, it can be used as an indicator of one particular side of the surface ∂D. The surface obtained by composition of $\mathbf{f}$ with $\mathbf{g}$ has the vector $\mathbf{Y_{f(g)}}$ as its side indicator and the surface is continuous without a boundary, just as ∂D is. If no point in D maps into $\mathbf{0}$, the previous argument showing that the integral (5.9) is zero applies again and $td(\mathbf{f}, D) = 0$. When one or more points of D are mapped into $\mathbf{0}$, then $td(\mathbf{f}, D)$ is still an integer, but this time the integer can be any value. To see this, imagine following a typical ray from $\mathbf{0}$ outward. The ray may intersect the surface $\mathbf{f}(\partial D)$ at a number of points, and at N_+ of these points the vector $\mathbf{Y}_{\mathbf{f}(\mathbf{g})}$ has a positive component in the direction of the ray, and at N_- of the points the vector has a negative component. On the sphere S_n, in a neighborhood of the ray's intersection point, the integral's mapping onto S_n multiplies the area of the neighborhood by the factor $N_+ - N_-$. As the ray is varied in direction, this factor cannot change until the ray crosses a boundary of the surface $\mathbf{f}(\partial D)$, and then because the surface is continuous, there is no net change in $N_+ - N_-$, because each of the integers N_+ and N_- increases by one, or each decreases by one. On the sphere S_n, except for an inconsequential set of points where the corresponding ray from $\mathbf{0}$ is tangent to the surface $\mathbf{f}(\partial D)$ or where some partial derivative of $\mathbf{f}$ fails to be defined, the factor $N_+ - N_-$ has only one value, and this integer value becomes $td(\mathbf{f}, D)$.

As an example in two-dimensional space, let D be a circle of radius a centered at the origin. We will take $\mathbf{f}$ as various functions defined over D. To compute $td(\mathbf{f}, D)$ for any $\mathbf{f}$, a mapping $\mathbf{g}$ is needed from u_1 space onto ∂D with the vector $\mathbf{Y}_{\mathbf{g}}$ pointing outward from ∂D. We may take $\mathbf{g}$ to be defined by the equations $g_1(u_1) = a \cos u_1$ and $g_2(u_1) = a \sin u_1$, with u_1 in $[0, 2\pi]$. If $\mathbf{f} = \mathbf{x}$, then we have

$$td(\mathbf{f}, D) = \frac{1}{2\pi} \int_0^{2\pi} \frac{1}{a^2} \det \begin{bmatrix} a \cos u_1 & a \sin u_1 \\ -a \sin u_1 & a \cos u_1 \end{bmatrix} du_1 = \frac{2\pi}{2\pi} = 1$$

If $\mathbf{f}$ is defined by $f_1 = -x_1, f_2 = x_2$, then we have

$$td(\mathbf{f}, D) = \frac{1}{2\pi} \int_0^{2\pi} \frac{1}{a^2} \det \begin{bmatrix} -a \cos u_1 & a \sin u_1 \\ a \sin u_1 & a \cos u_1 \end{bmatrix} du_1 = \frac{-2\pi}{2\pi} = -1$$

Finally, if $\mathbf{f}$ is defined by $f_1 = x_1^2 - x_2^2, f_2 = 2x_1 x_2$, we have

$$td(\mathbf{f}, D) = \frac{1}{2\pi} \int_0^{2\pi} \frac{1}{a^4} \det \begin{bmatrix} a^2 \cos 2u_1 & a^2 \sin 2u_1 \\ -2a^2 \sin 2u_1 & 2a^2 \cos 2u_1 \end{bmatrix} du_1 = \frac{4\pi}{2\pi} = 2$$

Suppose now that D is a box, and over this box a computed Jacobian interval for $\mathbf{f}$ is positive or negative. From the Corollary to Theorem 5.1, we know that $\mathbf{f}$ maps D one-to-one, so the surface $\mathbf{f}(\partial D)$ cannot intersect itself. The $td(\mathbf{f}, D)$ integral over $\mathbf{f}(\partial D)$ is just like the integral 5.9 over ∂D, so we may expect that $td(\mathbf{f}, D)$ equals 0 or 1 according as $\mathbf{0}$ is or is not in D. However, a specification for the integral 5.9 and for $td(\mathbf{f}, D)$ is that the vector $\mathbf{Y}_{\mathbf{g}}$ point outward from ∂D. The vector $\mathbf{Y}_{\mathbf{f}(\mathbf{g})}$ points outward from the $\mathbf{f}$ image of D if the Jacobian is positive. If the Jacobian is negative, orientation is reversed and this vector points inward, and this causes the integral to equal -1 if $\mathbf{0}$ is in D. For the special case of a negative Jacobian function $\mathbf{f}$ obtained by setting $f_i = x_i$ except for f_1 which equals $-x_1$, this result can be seen directly from 5.10, because then $td(\mathbf{f}, D) = -td(\mathbf{x}, D)$ and $td(\mathbf{x}, D)$ is the integral 5.9.

The topological degree has the property that if we divide the region D into two subregions D_1 and D_2, using some surface S_0 to effect the division, with S_0 understood to be a part of both D_1 and D_2, then

$$td(\mathbf{f}, D) = td(\mathbf{f}, D_1) + td(\mathbf{f}, D_2) \tag{5.11}$$

This follows because the two surface integrals over S_0 on the right side of the equation above cancel, because the respective $\mathbf{Y}$ vectors must point in opposite directions, and the sum of all other integrals over other surfaces on the right side equals the sum of the surface integrals on the left side.

Suppose there are a finite number k of isolated simple zeros in D. Using various dividing surfaces, we can divide D in subregions such that k of them, $D_1, \ldots, D_k$, each contain a single zero, the remaining subregions having no zero. Because the Jacobian is nonzero at each of the k zeros, we can arrange the subdivision so the regions D_i are small enough that the Jacobian throughout D_i has the same sign as the Jacobian at the contained zero. Employing the result 5.11, we see that when all zeros are isolated simple zeros, the topological degree equals the number of zeros with a positive Jacobian minus the number of zeros with a negative Jacobian, or, equivalently,

$$td(\mathbf{f}, D) = \sum_{\mathbf{x} \text{ a zero of } \mathbf{f}} \text{sgn}(\det J(\mathbf{x})) \tag{5.12}$$

The interesting properties of $td(\mathbf{f}, D)$ make it worthwhile to try to extend the definition of topological degree to include the case $n = 1$ in which a single function $f(x)$ is defined on some interval $[a, b]$. Here the boundary of D consists of the two points a and b, and it is not possible to define a surface integral over ∂D. However, if we take $A(S_1)$ as equal to 2, and replace the surface integral by a summation, an examination of 5.9 and 5.10 suggests

$$td(f, [a, b]) = \frac{1}{2} \left(\frac{f(a)}{|f(a)|} (-1) + \frac{f(b)}{|f(b)|} (+1) \right)$$

$$td(f, [a, b]) = \frac{1}{2} (\text{sgn } f(b) - \text{sgn } f(a)) \tag{5.13}$$

This is a valid extension, because if the signs of $f(a)$ and $f(b)$ are the same, the topological degree is the integer zero, otherwise it is either $+1$ or -1, and the nonzero value implies a zero in $[a, b]$. With appropriate reinterpretations, the relations 5.11 and 5.12 are still true for this case $n = 1$ topological degree extension.

5.7. COMPUTING THE TOPOLOGICAL DEGREE OVER A BOX B

The topological degree of an elementary function **f** over a box B can be calculated using interval arithmetic. The method may be viewed as an attempt to find $td(\mathbf{f}, B)$ by choosing a ray H from the origin and finding all the points where this ray crosses $\mathbf{f}(\partial B)$. If there are N_+ crossing points where the projected surface element gets a positive multiplier and N_- crossing points where the projected surface element gets a negative multiplier, then $N_+ - N_-$ is the topological degree.

To carry out the ray test procedure, we use a task queue holding domain that makes up the the boundary of B with each queue domain defined by fields s_i, $i = 1, \ldots, n$, previously used in Section 5.5 to test ∂B to make certain $\mathbf{f}(\mathbf{x}) \neq \mathbf{0}$ there. Recall that a field is flagged as a *point* holding a single number a or as an *interval* holding a number pair a, b, with $a < b$, and we distinguish the two cases by writing $s_i = a$ or $s_i = [a, b]$. Initially the queue is loaded with $2n$ domains designating the different sides of the box B.

A queue domain has one additional field, *orientation*, which equals either $+1$ or -1 and serves the following function. Consider the queue domain designating the side of B with x_j fixed at b_j. The vector $\mathbf{Y_g}$ must point outward, and so must be a positive multiple of $\mathbf{e}_j$. The integration variables $u_1, \ldots, u_{n-1}$ of $td(\mathbf{f}, B)$ over a side, are taken to be the variables $x_1 \ldots, x_{j-1}, x_{j+1}, \ldots, x_n$ associated with the $n - 1$ interval fields s_i. If these variables in the order given lead to the vector $\mathbf{Y_g}$ pointing inward instead of outward, this is corrected by setting the orientation field to -1. Otherwise the orientation field is $+1$. With our choice of integration variables, we have $g_i = x_i = u_i$ for $i = 1, \ldots, j - 1$, $g_j = b_j$, and $g_i = x_i = u_{i-1}$ for $i = j + 1, \ldots, n$. For the determinant of line 5.7 defining $\mathbf{Y_g}$, we see that column j has all zero elements except for the vector $\mathbf{e}_j$. A cofactor expansion of the determinant along column j gives us $(-1)^{j+1}\mathbf{e}_j$ times the determinant of an $(n - 1)$-square identity matrix. Thus $\mathbf{Y_g}$ equals $(-1)^{j+1}\mathbf{e}_j$, so the orientation field equals $(-1)^{j+1}$ to correct this. Similarly, on the side of B with $x_j = a_j$, the orientation field must be set to $(-1)^j$.

We take the ray H to be pointing in the positive direction of the first coordinate x_1. We search for regions of ∂B where the ray H would meet $\mathbf{f}(\partial B)$, the image of ∂B. We do this with the following task queue cycle: Taking the first queue domain, we set the primitives x_i equal to the corresponding s_i point or interval values. That is, one variable has a wid of zero, and the others have positive wids. Then using formal interval arithmetic, we generate interval values for the component functions $f_2, f_3, \ldots, f_n$. If any of these intervals do not contain 0, the ray H cannot intersect the image of this part of ∂B, and the queue domain is discarded. If all these function intervals contain 0, then we form an interval for f_1. If this interval is negative, again the ray H cannot meet the image of this part of ∂B, and the queue domain is discarded. If the f_1 interval is positive, it is likely that the ray H meets the image of this part of ∂B, and the queue domain is moved to another list L, initially empty, for later attention. In the remaining case, where the f_1 interval contains 0, it is uncertain whether or not the ray H meets the image of this part of the boundary of B, and we proceed as follows: From the queue domain we construct all possible queue domains that can be formed by replacing interval s_i fields by either their right half or their left half subintervals, copying point fields s_i, and copying the orientation value. These 2^{n-1} queue domains replace the first queue domain. (An alternate plan would be to form only two new domains by subdividing only the widest s_i interval, with these two domains replacing the first domain of the task queue.)

Because the point $\mathbf{0}$ does not lie in the image of ∂B, eventually the task queue becomes empty, although it may be necessary to increase the precision of computation to obtain this. (Every time a queue domain is subdivided, this is an appropriate time to check the precision of computation.) When the task queue is empty, the list L is examined. In general L contains domains that together define a subset $B^{(1)}$ of ∂B. The subset $B^{(1)}$ in general has a number of connected parts. On the boundary of each connected part the functions $f_2, f_3, \ldots, f_n$ can never be simultaneously zero. This is because each point on the boundary of the connected part is also a point on the boundary of some queue domain that was discarded. This discarded queue domain had one of its functions $f_2, \ldots, f_n$ positive or negative or had f_1 negative, but f_1 negative did not occur because f_1 was positive on the adjoining connected part.

Consider any domain on L with orientation σ defining a subdomain S of $B^{(1)}$. The ray H may meet the $\mathbf{f}$ image of S at one or more points where $\mathbf{f}(\mathbf{x})$ is a positive multiple of $\mathbf{e}_1$, and for each of these points there is a point P of S where the functions $f_2, f_3 \ldots, f_n$ are simultaneously zero. According to Equation 5.10 the surface element projected onto the sphere S_n has a positive or negative multiplier according as

$$\sigma \det \begin{bmatrix} \dfrac{\partial f_2}{\partial u_1} & \dfrac{\partial f_3}{\partial u_1} & \cdots & \dfrac{\partial f_n}{\partial u_1} \\[2mm] \dfrac{\partial f_2}{\partial u_2} & \dfrac{\partial f_3}{\partial u_2} & \cdots & \dfrac{\partial f_n}{\partial u_2} \\[2mm] \vdots & \vdots & \vdots\vdots\vdots & \vdots \\[2mm] \dfrac{\partial f_2}{\partial u_{n-1}} & \dfrac{\partial f_3}{\partial u_{n-1}} & \cdots & \dfrac{\partial f_n}{\partial u_{n-1}} \end{bmatrix}$$

is positive or negative. By the relation 5.12 the contribution made by all such points to the sum $N_+ - N_-$ is σ times the topological degree of the R^{n-1} to

R^{n-1} mapping $\mathbf{f}^{(1)}$ defined by $f_2, f_3, \ldots, f_n$ on the $(n-1)$-dimensional region S (embedded in R^n). We need to sum the contribution $\sigma\ td(\mathbf{f}^{(1)}, S)$ for each domain on the list L to obtain $N_+ - N_-$. Thus we have replaced our original task of calculating a topological degree of a function with n components, by the task of calculating a set of signed topological degrees of a function with $n-1$ components.

To do this second task, we proceed much as we did before, this time loading our task queue with boundary domains for each element on the list L. For each domain of L with orientation σ, we form $2^{(n-1)}$ boundary domains for the task queue by copying all fields s_i except for one interval field that is converted to a point field, being set to an endpoint of the interval. If the interval field converted was the jth interval field, counting from s_1 toward s_n, then the orientation assigned is σ times $(-1)^j$ or times $(-1)^{j+1}$ according as the left or the right endpoint is assigned. If two L domains share a side, then this side gets specified twice on the queue. Such pairs are discarded, because the orientation values assigned to them are always opposite, and so their net contribution to the sum being formed is zero. It is also possible, because of the domain subdivision process sometimes employed, for an L domain's side to be contained within another L domain's side. In this case the task queue is corrected to specify only the unrepeated part of the larger side. After such duplications are removed, what is left on the queue defines the boundary of $B^{(1)}$. Note that the formation of the initial task queue is similar to the formation of this second task queue, if we imagine an initial list L containing just B with an orientation of $+1$.

The second task queue computation is similar to the first, except that the number of f_i functions evaluated is reduced by one. This time we take the ray H to be pointing in the positive direction of the second coordinate x_2. Eventually the task queue is empty and there is another list L specifying a subdomain $B^{(2)}$ of $\partial B^{(1)}$ that requires further examination.

Each time the process is carried out, the dimension of the point sets examined and the number of functions treated is reduced by one. Just before the n-th iteration, the list L defines a set $B^{(n-1)}$, the list containing a number of line segments S imbedded in R^n, each with an orientation σ assigned. The sum of contributions $\sigma\ td(\mathbf{f}^{(n-1)}, S)$ must be found to obtain $N_+ - N_-$. Now $\mathbf{f}^{(n-1)}$ has just one component $f_n(\mathbf{x})$, so its topological degree no longer is defined with a surface integral but by Equation 5.13. But if we follow the system we have used on earlier cycles, we find that this leads to a correct topological degree evaluation in this case also. That is, for each line segment S with orientation σ on the list L, we construct two point domain elements for the task queue, the orientation value assigned being $+\sigma$ for the point domain having the right endpoint of the single remaining S interval field and $-\sigma$ for the other. The ray H now points in the positive direction of x_n, and the final queue cycle is to move to the list L only those point domains $\mathbf{x}$, which have $f_n(\mathbf{x})$ positive, the others being discarded.

After this final task queue is empty, the sum of the orientations of the point domains on L yields $N_+ - N_-$, and we finally obtain the topological degree of $\mathbf{f}(\mathbf{x})$ over B.

An example may clarify the general process. Consider the mapping from R^3 into R^3 defined by

$$f_1 = x_1^2 + x_2^2 - x_3, \qquad f_2 = x_2^2 + x_3^2 - x_1, \qquad f_3 = x_3^2 + x_1^2 - x_2$$

These functions have a zero at $(0, 0, 0)$ and at $(\frac{1}{2}, \frac{1}{2}, \frac{1}{2})$. The Jacobian at $(0, 0, 0)$ is

$$\det \begin{bmatrix} 0 & 0 & -1 \\ -1 & 0 & 0 \\ 0 & -1 & 0 \end{bmatrix} = -1$$

and the Jacobian at $(\frac{1}{2}, \frac{1}{2}, \frac{1}{2})$ is

$$\det \begin{bmatrix} 1 & 1 & -1 \\ -1 & 1 & 1 \\ 1 & -1 & 1 \end{bmatrix} = 4$$

Thus $td(\mathbf{f}, B)$ equals -1 if B is a small box containing just the zero $(0, 0, 0)$, equals $+1$ if B is a small box containing just the zero $(\frac{1}{2}, \frac{1}{2}, \frac{1}{2})$, and equals zero if B is a box large enough to contain both zeros. We compute the degree for a box B containing just the origin, namely the box B with the interval $[-\frac{1}{4}, \frac{1}{4}]$ for each variable x_i.

There are initially six domains defining sides of B on the first task queue, and only the domain given below is moved to the list L when it is tested.

$$s_1 = [-\frac{1}{4}, \frac{1}{4}], \qquad s_2 = [-\frac{1}{4}, \frac{1}{4}], \qquad s_3 - -\frac{1}{4}; \qquad \sigma = -1 \quad (5.14)$$

For this domain, the interval for both f_2 and f_3 is $[-\frac{3}{16}, \frac{3}{8}]$, containing 0 as required, and the interval for f_1 is $[\frac{1}{4}, \frac{3}{8}]$, positive as required. All of the other five L domains are discarded, either because their f_2 interval does not contain 0, their f_3 interval does not contain 0, or because their f_1 interval is negative.

The region $B^{(1)}$ is defined by the single domain 5.14, and so the queue on the second iteration has four domains defining its sides. Of these only the one given below is transferred to L when it is tested.

$$s_1 = -\frac{1}{4}, \qquad s_2 = [-\frac{1}{4}, \frac{1}{4}], \qquad s_3 = -\frac{1}{4}; \qquad \sigma = 1 \quad (5.15)$$

For this domain, the interval for f_3 is $[-\frac{1}{8}, \frac{3}{8}]$, containing 0 as required, and the interval for f_2 is $[\frac{5}{16}, \frac{3}{8}]$, positive as required. The other three domains either have an f_3 interval that does not contain 0 or have a negative f_2 interval.

The region $B^{(2)}$ is defined by the single domain 5.15, and so the list L on the last iteration has two domains defining its sides, which now are points. Only the domain below yields a positive sign for f_3.

$$s_1 = -\frac{1}{4}, \qquad s_2 = -\frac{1}{4}, \qquad s_3 = -\frac{1}{4}; \qquad \sigma = -1$$

Accordingly the degree equals -1, the orientation value of this single domain.

5.8. NEWTON'S METHOD FOR THE GENERAL PROBLEM

Suppose the n component functions $f_i(\mathbf{x}) = f_i(x_1, x_2, \ldots, x_n)$ making up $\mathbf{f}(\mathbf{x})$ and all partial derivatives $\frac{\partial f_i(\mathbf{x})}{\partial x_j}$ are defined in a certain region of $\mathbf{x}$ space, and suppose that $\mathbf{x}^{(k)} = (x_1^{(k)}, x_2^{(k)}, \ldots, x_n^{(k)})$ is a zero approximation for $\mathbf{f}(\mathbf{x})$. The Taylor series of the component functions, with the series expansion point $\mathbf{x}^{(k)}$, are

$$f_1(\mathbf{x}) = f_1(\mathbf{x}^{(k)}) + \sum_{j=1}^{n} \frac{\partial f_1(\mathbf{x}^{(k)})}{\partial x_j}(x_j - x_j^{(k)}) + \cdots$$

$$f_2(\mathbf{x}) = f_2(\mathbf{x}^{(k)}) + \sum_{j=1}^{n} \frac{\partial f_2(\mathbf{x}^{(k)})}{\partial x_j}(x_j - x_j^{(k)}) + \cdots$$

$$\vdots \qquad \vdots \qquad\qquad \vdots$$

$$f_n(\mathbf{x}) = f_n(\mathbf{x}^{(k)}) + \sum_{j=1}^{n} \frac{\partial f_n(\mathbf{x}^{(k)})}{\partial x_j}(x_j - x_j^{(k)}) + \cdots$$

A linear approximation to each function f_i can be obtained from just the series terms shown above. If $J(\mathbf{x}^{(k)})$ is the matrix with i, j element $\frac{\partial f_i(\mathbf{x}^{(k)})}{\partial x_j}$, then the linear approximations may be written in matrix–vector form as

$$\mathbf{y} = \mathbf{f}(\mathbf{x}^{(k)}) + J(\mathbf{x}^{(k)})(\mathbf{x} - \mathbf{x}^{(k)})$$

If we set $\mathbf{y}$ equal to $\mathbf{0}$, and solve for $\mathbf{x}$ to obtain a better zero approximation, calling our solution $\mathbf{x}^{(k+1)}$, we obtain the equation

$$\mathbf{x}^{(k+1)} = \mathbf{x}^{(k)} - [J(\mathbf{x}^{(k)})]^{-1}\mathbf{f}(\mathbf{x}^{(k)}) \tag{5.16}$$

In Section 5.5 the method described there bounds a single simple zero inside a box $B^{(C)}$ with the $\mathbf{f}$ Jacobian known to be nonzero over the box. Using the iteration equation (5.16) and with the starting approximation $\mathbf{x}^{(0)}$ taken as the box centerpoint, the sequence of iterates generally rapidly locates the zero. After an iterate $\mathbf{x}^{(k+1)}$ is computed, it is compared with $\mathbf{x}^{(k)}$, and if for some i we find $x_i^{(k+1)} \neq x_i^{(k)}$, then, in preparation for the next cycle, $\mathbf{x}^{(k+1)}$ is replaced by mid

$\mathbf{x}^{(k+1)}$, that is, the vector whose i-th component is mid $x_i^{(k+1)}$. Eventually we find $x_i^{(k+1)} \doteq x_i^{(k)}$ for all i, and then mid $\mathbf{x}^{(k)}$ defines the final zero approximation $\mathbf{x}^{(\mathbf{z})}$.

An error bound for $\mathbf{x}^{(\mathbf{z})}$ can be obtained by generalizing the error bounding method used in Section 5.3. If $\mathbf{z}$ is the true zero inside $B^{(C)}$, then $\mathbf{f}(\mathbf{z}) = \mathbf{0}$, and using the Mean value theorem once more, we have for $i = 1, \ldots, n$,

$$f_i(\mathbf{x}^{(\mathbf{z})}) = f_i(\mathbf{x}^{(\mathbf{z})}) - f_i(\mathbf{z}) = \sum_{j=1}^{n} \frac{\partial f_i(\mathbf{c}_{ij})}{\partial x_j} (x_j^{(\mathbf{z})} - z_j)$$

where the points $\mathbf{c}_{ij}$ lie on the line segment joining $\mathbf{x}^{(\mathbf{z})}$ and $\mathbf{z}$. If $\hat{J}$ is the n-square matrix with elements $\frac{\partial f_i(\mathbf{c}_{ij})}{\partial x_j}$, then we have

$$\mathbf{f}(\mathbf{x}^{(\mathbf{z})}) = \hat{J}(\mathbf{x}^{(\mathbf{z})} - \mathbf{z})$$

which implies

$$\mathbf{x}^{(\mathbf{z})} - \mathbf{z} = \hat{J}^{-1}\mathbf{f}(\mathbf{x}^{(\mathbf{z})})$$

Each element of $\hat{J}$ is contained in the corresponding interval element of the interval matrix $J(\mathbf{x}^{(C)}) \pm W$, which is formed in the process of testing the container box $B^{(C)}$. So if an interval inverse matrix $[J(\mathbf{x}^{(C)}) \pm W]^{-1}$ is formed, each component of $\mathbf{x}^{(\mathbf{z})} - \mathbf{z}$ is contained within the interval elements of $\mathbf{y} = [J(\mathbf{x}^{(C)}) \pm W]^{-1} \mathbf{f}(\mathbf{x}^{(\mathbf{z})})$. Thus the error of $x_i^{(\mathbf{z})}$ is bounded by $e_i = |\text{mid } y_i| + \text{wid } y_i$. Then $x_i^{(\mathbf{z})} \oplus e_i$ is a correctly ranged zero approximation component. If our error bounding process fails to yield a zero with sufficient correct decimal places, then it is necessary to increase the precision of computation, restart the Newton's method iteration, and repeat this test with a better $\mathbf{x}^{(\mathbf{z})}$ approximation.

5.9. SEARCHING A MORE GENERAL REGION FOR ZEROS

Suppose $\mathbf{f}$ has two components, and we are interested in finding all the zeros of $\mathbf{f}$ that lie inside the unit circle $x_1^2 + x_2^2 \le 1$. Solvable Problem 5.4 requires the search domain to be a rectangle, and we could enclose the unit circle in a square S and search within S, but there is a difficulty that could arise. It is possible that a component function f_i is defined inside the unit circle but not in the square.

It is worthwhile to consider now whether Solvable Problem 5.4 or 5.5 can be changed to allow more general search domains than just boxes. An iterated integral for n dimensional volume of the form

$$\int_a^b \int_{g_1(x_1)}^{h_1(x_1)} \int_{g_2(x_1,x_2)}^{h_2(x_1,x_2)} \cdots \int_{g_{n-1}(x_1,x_2,\ldots,x_{n-1})}^{h_{n-1}(x_1,x_2,\ldots,x_{n-1})} dx_n dx_{n-1} \cdots dx_2 dx_1 \quad (5.17)$$

has an integration domain that can serve also as a zero search domain. For instance, for our example problem of finding zeros within the unit circle, we could use the dimension 2 domain of

$$\int_{-1}^{1} \int_{-\sqrt{1-x_1^2}}^{\sqrt{1-x_1^2}} dx_2 dx_1$$

To solve Problem 5.5 over boxes, we employed a mapping from a reference unit box $B^{(u)}$ into the problem box B. For the circle example we can map from the unit square into the circle with a function $\mathbf{x} = \mathbf{v}(\mathbf{u})$ defined by

$$x_1 = -1 \cdot (1 - u_1) + 1 \cdot u_1$$

$$x_2 = -\sqrt{1 - x_1^2} \cdot (1 - u_2) + \sqrt{1 - x_1^2} \cdot u_2$$

Here u_1 and u_2 vary in $[0, 1]$. This mapping allows us to treat a function $\mathbf{f}$ defined in the circle as if it were defined in the unit square. The mapping from the boundary of the unit square into the boundary of the circle is not one-to-one, because the entire left and right edges of the unit square map onto points, but the mapping from the interior of the unit square into the interior of the circle is one-to-one. Using the composite $\mathbf{f}$, we can search the interior of $B^{(u)}$ for zeros, mapping each zero in $B^{(u)}$ into a zero in the circle.

DEFINITION 5.1: An elementary region is a region of n-space that can defined as the integration domain of the integral (5.17) where all functions h_i and g_i are elementary functions.

An elementary region of dimension 1 is a closed interval $[a, b]$. An elementary region R of dimension 2 is defined by two elementary functions $h_1(x_1)$ and $g_1(x_1)$ over an interval $[a, b]$, and its interior consist of those points (x, y) satisfying $a < x < b$ and y between $g_1(x)$ and $h_1(x)$.

The method described for solving Problem 5.5 also serves for solving the more general problem given below.

SOLVABLE PROBLEM 5.6: For any elementary function $\mathbf{f}(\mathbf{x})$ defined on an elementary region R, having only isolated zeros in R, for a specific positive integer m, locate to m decimals a point on the boundary of R where all components of $\mathbf{f}(\mathbf{x})$ are less in magnitude than 10^{-m} and halt, or, if $\mathbf{f}(\mathbf{x}) \neq \mathbf{0}$ on the boundary of R, bound all the zeros of $\mathbf{f}$ in R by giving to k decimals

1. Points identified as simple zeros, or
2. Points identified as containing within their tilde box a region where $\max_i |f_i(\mathbf{x})| < 10^{-k}$, the region identified as *containing at least one zero* if the $\mathbf{f}(\mathbf{x})$ topological degree over a bounding box is nonzero.

With an elementary region R, we map from a unit box $B^{(u)}$ into R using a mapping $\mathbf{x} = \mathbf{x}(\mathbf{u})$ derived from the functions defining R. Using the notation of the integral (5.17), we have $x_1 = (1 - u_1)a + u_1 b$, and, for $i > 1$, $x_i = (1 - u_i)g_{i-1} + u_i h_{i-1}$. The mapping from the boundary of $B^{(u)}$ into the boundary of R may or may not be one-to-one, but the mapping from the interior of $B^{(u)}$ to the interior of R is one-to-one, so we can search for zeros of $\mathbf{f}$ in the interior of R by searching for zeros of $\mathbf{f}(\mathbf{x}(\mathbf{u}))$ in the interior of $B^{(u)}$.

The zero finding process that we used previously also will work for the more general region R. The $B^{(u)}$ box is subdivided just as before, but the determination of a degree 0 or degree 1 series expansion for $\mathbf{f}$ now is done in two steps reflecting $\mathbf{f}$'s composite nature. For any $\mathbf{u}$ subbox $B^{(s)}$ of $B^{(u)}$, or any container box $B^{(c)}$, we obtain an $\mathbf{x}$ subbox $B^{(x)}$ by obtaining interval values for the various x_i expressions given in the preceding paragraph. Then after x_i primitives are set accordingly, we obtain the f_i expansions.

As before, we test $\mathbf{f}$ to make certain it is nonzero over the boundary of $B^{(u)}$. The search for zeros is done with a task queue listing subboxes of $B^{(u)}$ in which a zero is possible, but any accuracy tests are with the derived $B^{(x)}$ subboxes.

EXERCISES

1. Call `findzero` and find to 10 decimals the zeros of

$$f_1(x_1, x_2) = x_1^2 - x_2$$
$$f_2(x_1, x_2) = x_2^2 - x_1$$

Use the search interval $[-10, 10]$ for both x_1 and x_2. Obtain in this way the two solutions $x_1 = x_2 = 0$ and $x_1 = x_2 = 1$.

2. Call `findzero` using the problem of Exercise 1 but with the search interval $[1, 10]$ for both x_1 and x_2. Note the program's rejection of the problem because the functions are not determined to be nonzero on the boundary of the search region.

3. Call `findzero` using the problem of Exercise 1 but with the search region a circle centered at the origin, using first a radius of 2, then a radius of $\sqrt{2}$, and finally a radius of $\frac{1}{2}$. Note the various responses of the program.

4. Call `findzero` and find to 10 decimals the zeros of $x^{1/3}$ in $[-5, 5]$. (To enter a fractional exponent p/q, the keystroke sequence is `^ (p/q)`.) Here there is a single zero, which is not simple, because the derivative at the zero is undefined.

5. Call findzero and find to 10 decimals the zeros of $x^2 + 10^{-100}$ in $[-1, 1]$. Here there is no zero, but findzero locates a zero possibility. Increase the number of decimals requested to 50, and note that this time no zero is found.

6. Call findzero and find to 10 decimals the zeros of the three component function used as an example in Section 5.7:

$$f_1 = x_1^2 + x_2^2 - x_3, \qquad f_2 = x_2^2 + x_3^2 - x_1, \qquad f_3 = x_3^2 + x_1^2 - x_2$$

Take the search interval for each component to be $[-10, 10]$.

7. Call findzero and find to one decimal the single zero of the two functions

$$f_1 = x_1^2 + x_2^2 - 2, \qquad f_2 = x_1 + x_2 - 2$$

Take the search interval for each component to be $[-10, 10]$. The straight line $x_1 + x_2 = 2$ is tangent to the circle $x_1^2 + x_2^2 - 2$ at the point $(1, 1)$, and the Jacobian is zero there, so this zero is found by the slow method of reducing the size of the container region. When zeros of this type are possible, it is best to request only a small number of correct decimal places.

8. Call findzero and find to 5 decimals the single zero of

$$f_1 = x_1^2 - x_2, \qquad f_2 = (x_1 - a)^2 - x_2$$

Set the parameter a initially to 0.1, and then repeat the problem with a equal to 0.01. The zero at $(\frac{a}{2}, \frac{a^2}{4})$ becomes more difficult to find as the parameter a approaches 0, and the problem can be made computer bound by making a small enough.

NOTES

N5.1. For an excellent discussion of interval versions of Newton's method, see the book by Hansen [4].

N5.2. The book by Cronin [3] develops the theory of the topological degree using concepts from combinatorial topology.

N5.3. The Kronecker integral for topological degree is given in the text by Alexandroff and Hopf [2, pp. 465–467].

N5.4. The method given in Section 5.7 for computing topological degree using interval arithmetic [1], is a version of a method proposed by Kearfott [7]. Two other methods of using Kronecker's integral to compute the topological degree are given by O'Neal and Thomas [9] and by Stenger [11].

N5.5. Other methods for finding zeros of several functions of several variables are given in Neumaier's book [8]. See also Harvey and Stenger [5], Kearfott [6], and Sikorski [10].

REFERENCES

1. Aberth, O., Computation of topological degree using interval arithmetic, and applications, *Math. of Computation* **62** (1994), 171–178.
2. Alexandroff P. and Hopf, H., *Topologie*, Chelsea, New York, 1935.
3. Cronin, J., *Fixed points and topological degree in nonlinear analysis*, Math Surveys, no. 11, American Mathematical Society, Providence, 1964.
4. Hansen, E., *Global Optimization using Interval Analysis*, Marcel Dekker, Inc., New York, 1992.
5. Harvey, C. and Stenger, F., A two-dimensional analogue to the method of bisections for solving nonlinear equations, *Quart. Appl. Math.* **33** (1975), 351–368.
6. Kearfott, R. B., An efficient degree-computation method for a generalized method of bisection, *Numer. Math.* **32** (1979), 109–127.
7. Kearfott, R. B., *A summary of recent experiments to compute the topological degree*, Proceedings of an International Conference on Applied Nonlinear Analysis, University of Texas at Arlington, April 20–22, 1978, (V. Laskshmikantham, ed.), Academic Press, New York, 1979, pp. 627–633.
8. Neumaier, A., *Interval methods for systems of equations, Encyclopedia of Mathematics and its Applications*, Cambridge University Press, Cambridge, 1990.
9. O'Neal, T. and Thomas, J., The calculation of the topological degree by quadrature, *SIAM J. Numer. Anal.* **12** (1975), 673–680.
10. Sikorski, K., A three-dimensional analogue to the method of bisections for solving nonlinear equations, *Math. Comp.* **33** (1979), 722–738.
11. Stenger, F., Computing the topological degree of a mapping in R^n, *Numer. Math.* **25** (1975), 23–38.

VI

FINDING ZEROS OF POLYNOMIALS AND OTHER ANALYTIC FUNCTIONS

6.1. POLYNOMIALS

A polynomial of degree n in the variable z, the integer n being nonnegative, has the form

$$P(z) = c_n z^n + c_{n-1} z^{n-1} + \cdots + c_1 z + c_0 \qquad (6.1)$$

where the leading coefficient c_n is required to be nonzero. The polynomial is *real, complex, rational*, or *complex rational*, according as all coefficients c_i are, respectively, real, complex, rational, or complex rational numbers. Any polynomial also has the representation

$$P(z) = c_n(z - z_1)(z - z_2) \cdots (z - z_n) \qquad (6.2)$$

where the n factors $z - z_i$ are unique, apart from order. The complex numbers z_i are zeros of the polynomial, and often these numbers are also called *roots* of the polynomial. Going from the representation 6.1 to the representation 6.2 is not a simple problem, but at least one that is solvable.

SOLVABLE PROBLEM 6.1: For any real or complex polynomial of positive degree n, find the n zeros to k decimals.

We find a complex number to k decimal places if we find both its real part and its imaginary part to k places. Efficient methods of locating the zeros of real or complex polynomials are described in Section 6.3.

Some of the zeros of a polynomial may be identical. If in Equation 6.2, we group together the linear factors of identical zeros and reassign indices so that the distinct zeros are $z_1, z_2, \ldots, z_q$, we obtain the representation

$$P(z) = c_n(z - z_1)^{n_1}(z - z_2)^{n_2} \cdots (z - z_q)^{n_q} \tag{6.3}$$

Here the integers $n_1, n_2, \ldots, n_q$ sum to n. The exponent n_i is the *multiplicity* of the zero z_i. If n_i is one, z_i is a *simple* zero, otherwise z_i is a *multiple* zero. It is not a solvable problem to find the multiplicity of the zeros of a real or complex polynomial of degree more than 1, as is easily seen by considering the polynomial

$$P(z) = z^2 - (a_1 + a_2)z + a_1 a_2 = (z - a_1)(z - a_2)$$

where a_1 and a_2 are any real numbers. If we can always determine the multiplicity of this polynomial's zeros, then we have a method of determining whether any two numbers a_1 and a_2 are equal, contradicting Nonsolvable Problem 3.1.

NONSOLVABLE PROBLEM 6.2: For any real or complex polynomial of degree $n > 1$, determine the multiplicity of the zeros.

Suppose for a polynomial $P(z)$ we obtain correctly ranged approximations to all its zeros, using the methods described in later sections. We can arrange the zeros in sets, with two zeros $z_1 = x_1 + iy_1$ and $z_2 = x_2 + iy_2$ belonging to the same set if we find $x_1 \doteq x_2$ and $y_1 \doteq y_2$. A zero that is in a set all by itself is certainly a simple zero, but suppose we find m zeros collected in one set. We then have a zero of *apparent* multiplicity m. If we were to compute polynomial zero approximations to more correct decimal places, the zeros may arrange themselves differently, so that a zero that was of apparent multiplicity m becomes a zero of smaller apparent multiplicity or even a simple zero. Thus an apparent multiplicity may be larger than the true multiplicity, and only an apparent multiplicity of one is certain to be the correct multiplicity.

For rational or complex rational polynomials this difficulty determining multiplicity disappears.

SOLVABLE PROBLEM 6.3: For any rational or complex rational polynomial of positive degree n, find its distinct zeros to k decimals, and determine their multiplicity.

From a rational or complex rational polynomial $P(z)$, we can obtain a set of polynomials $N_1(z), N_2(z), \ldots, N_q(z)$ containing the distinct zeros of $P(z)$, such that the zeros of $P(z)$ which are of multiplicity m are simple zeros of $N_m(z)$.

The task of finding the zeros of $P(z)$ is made easier by this decomposition, and the correct multiplicity of these zeros is obtained as a by-product.

A polynomial with a leading coefficient of 1 is called a *monic* polynomial. Any polynomial can be converted to monic form by dividing it by its leading coefficient c_n. If this is done for the polynomial 6.1, the polynomial becomes

$$P_1(z) = z^n + a_{n-1}z^{n-1} + \cdots + a_1 z + a_0$$

where the coefficient a_i equals c_i/c_n. The representation 6.3 for $P_1(z)$ is

$$P_1(z) = (z - z_1)^{n_1}(z - z_2)^{n_2} \cdots (z - z_q)^{n_q} \qquad (6.4)$$

If z_i is a zero of multiplicity n_i for $P_1(z)$, then z_i is a zero of multiplicity $n_i - 1$ for the derivative of $P_1(z)$. This is easy to show. We have

$$P_1(z) = (z - z_i)^{n_i}S(z)$$

where $S(z)$ is some polynomial of lower degree not having z_i as a zero, that is, $S(z_i) \neq 0$. Taking the derivative, we have

$$P_1'(z) = n_i(z - z_i)^{n_i-1}S(z) + (z - z_i)^{n_i}S'(z) = (z - z_i)^{n_i-1}[n_iS(z) + (z - z_i)S'(z)]$$

$$= (z - z_i)^{n_i-1}T(z)$$

We see that the multiplicity of z_i for $P_1'(z)$ is at least $n_i - 1$, and it cannot be higher because $T(z_i) = n_iS(z_i) \neq 0$. Thus if the multiplicity structure of $P_1(z)$ is as shown in Equation 6.4, then $P_1(z)$ and $P_1'(z)$ have the common divisor

$$P_2(z) = (z - z_1)^{n_1-1}(z - z_2)^{n_2-1} \cdots (z - z_q)^{n_q-1} \qquad (6.5)$$

and cannot have any common divisor of higher degree, so $P_2(z)$ may be considered a greatest common divisor. Any two polynomials have a unique monic greatest common divisor polynomial, as is clear by considering the linear factor forms 6.3 of the two polynomials. If $P_1(z)$ has rational or complex rational coefficients, we can find the greatest common divisor of $P_1(z)$ and $P_1'(z)$ by using a version of the Euclidean algorithm described in Section 2.7. The polynomial $P_1'(z)$ is divided into $P_1(z)$ using the usual polynomial division procedure, obtaining the quotient polynomial $Q_1(z)$ and the remainder polynomial $R_1(z)$:

$$P_1(z) = Q_1(z)P_1'(z) + R_1(z)$$

The pair $P_1(z)$, $P_1'(z)$ has the same greatest common polynomial divisor as the pair $P_1'(z)$, $R_1(z)$, because any common divisor polynomial of $P_1(z)$ and $P_1'(z)$ is also a common divisor of $R_1(z)$, and any common divisor polynomial of $P_1'(z)$

and $R_1(z)$ is also a common divisor of $P_1(z)$. The division process is repeated for the new pair, obtaining

$$P_1'(z) = Q_2(z)R_1(z) + R_2(z)$$

and this process is continued until a zero remainder is obtained:

$$R_1(z) = Q_3(z)R_2(z) \quad + R_3(z)$$
$$R_2(z) = Q_4(z)R_3(z) \quad + R_4(z)$$
$$\vdots \qquad \vdots \qquad\qquad \vdots$$
$$R_{s-2}(z) = Q_s(z)R_{s-1}(z) + R_s(z)$$
$$R_{s-1}(z) = Q_{s+1}(z)R_s(z)$$

The polynomial $R_s(z)$, after it is converted to monic form, equals $P_2(z)$ of Equation 6.5.

The procedure used on $P_1(z)$ to obtain $P_2(z)$ can now be applied to $P_2(z)$ if this monic polynomial is not equal to 1, and then we get the polynomial $P_3(z)$. Thus the Euclidean procedure, if repeatedly applied, produces a series of monic polynomials $P_1(z), P_2(z), P_3(z), \ldots, P_q(z), 1$ that always ends with a polynomial $P_{q+1}(z)$ equal to 1.

When $P_1(z)$ of line 6.4 is divided by the polynomial $P_2(z)$ of line 6.5, we obtain the polynomial

$$M_1(z) = (z - z_1)(z - z_2) \cdots (z - z_q)$$

with all the distinct zeros of $P_1(z)$ as simple zeros. In general, if for the sequence $P_1(z), P_2(z), \ldots, P_q(z), P_{q+1}(z) = 1$, we define

$$M_i(z) = \begin{cases} \dfrac{P_i(z)}{P_{i+1}(z)} & \text{if } i < q \\ P_q(z) & \text{if } i = q \end{cases}$$

we get the set of polynomials $M_1(z), M_2(z), \ldots, M_q(z)$, all with only simple zeros, such that all zeros of $P_1(z)$, which are of multiplicity m or higher, are zeros of $M_m(z)$. From these polynomials it is convenient to obtain by division another set of q polynomials $N_1(z), N_2(z), \ldots, N_q(z)$, again with simple zeros only, defined by the equation

$$N_i(z) = \begin{cases} \dfrac{M_i(z)}{M_{i+1}(z)} & \text{if } i < q \\ M_q(z) & \text{if } i = q \end{cases}$$

The polynomials $N_1(z)$ have degrees that sum to q, the number of distinct zeros of $P_1(z)$, and each distinct zero of $P_1(z)$ is a zero of just one of these polynomials. If the zero is of multiplicity m, then it is a zero of $N_m(z)$. Thus when the zeros of the polynomials $N_i(z)$ are found by the methods described in Section 6.3, they can be displayed with their exact multiplicity.

As an example, suppose $P_1(z) = z^7 - 3z^5 + 3z^3 - z$. Then $P'_1(z) = 7z^6 - 15z^4 + 9z^2 - 1$.

The Euclidean algorithm yields

$$z^7 - 3z^5 + 3z^3 - z = \frac{1}{7}z(7z^6 - 15z^4 + 9z^2 - 1) - \frac{6}{7}z^5 + \frac{12}{7}z^3 - \frac{6}{7}z$$

$$7z^6 - 15z^4 + 9z^2 - 1 = -\frac{49}{6}z(-\frac{6}{7}z^5 + \frac{12}{7}z^3 - \frac{6}{7}z) - z^4 + 2z^2 - 1$$

$$-\frac{6}{7}z^5 + \frac{12}{7}z^3 - \frac{6}{7}z = \frac{6}{7}z(-z^4 + 2z^2 - 1)$$

Making the last polynomial of the series monic by dividing by its leading coefficient, we get $P_2(z) = z^4 - 2z^2 + 1$. Repeating the Euclidean algorithm with $P_2(z)$, we obtain

$$z^4 - 2z^2 + 1 = \frac{1}{4}z(4z^3 - 4z) - z^2 + 1$$

$$4z^3 - 4z = -4z(-z^2 + 1)$$

so $P_3(z) = z^2 - 1$. One more application of the Euclidean algorithm leads to

$$z^2 - 1 = \frac{1}{2}z(2z) - 1$$

$$2z = -2z(-1)$$

so $P_4(z)$ is 1.

The $M_i(z)$ polynomials are

$$M_1(z) = \frac{P_1(z)}{P_2(z)} = z^3 - z$$

$$M_2(z) = \frac{P_2(z)}{P_3(z)} = z^2 - 1$$

$$M_3(z) = P_3(z) = z^2 - 1$$

and the $N_i(z)$ polynomials are

$$N_1(z) = \frac{M_1(z)}{M_2(z)} = z$$

$$N_2(z) = \frac{M_2(z)}{M_3(z)} = 1$$

$$N_3(z) = M_3(z) = z^2 - 1$$

We see that the zeros of $P_1(z)$ are 0 of multiplicity 1, and 1, -1, both of multiplicity 3.

These results can also be understood by setting

$$P_1(z) = (z - z_1)^3(z - z_2)^3(z - z_3)$$

with $z_1 = 1$, $z_2 = -1$, and $z_3 = 0$. We see that by repeated application of the Euclidean algorithm, we would obtain the sequence

$$P_2(z) = (z - z_1)^2(z - z_2)^2$$

$$P_3(z) = (z - z_1)(z - z_2)$$

$$P_4(z) = 1$$

leading to the sequence

$$M_1(z) = (z - z_1)(z - z_2)(z - z_3)$$

$$M_2(z) = (z - z_1)(z - z_2)$$

$$M_3(z) = (z - z_1)(z - z_2)$$

and then finally to the sequence

$$N_1(z) = (z - z_3)$$

$$N_2(z) = 1$$

$$N_3(z) = (z - z_1)(z - z_2)$$

6.2. A BOUND FOR THE ZEROS OF A POLYNOMIAL

When searching for the zeros of a monic polynomial

$$P_1(z) = z^n + a_{n-1}z^{n-1} + \cdots + a_1z + a_0 \tag{6.6}$$

it is helpful to compute a *bounding radius* R such that in the complex plane the circle $|z| \le R$ contains all the zeros. The theorems below give two ways of computing a bounding radius.

THEOREM 6.1: *A bounding radius for the zeros of the polynomial $P_1(z)$ is*

$$R_1 = \max\{|a_{n-1}| + 1, |a_{n-2}| + 1, \ldots, |a_1| + 1, |a_0|\}$$

As an example, let us form the polynomial

$$P_1(z) = (z - 1)(z - 2)(z - 3)(z - 4) = z^4 - 10z^3 + 35z^2 - 50z + 24$$

We obtain $R_1 = \max [11, 36, 51, 24] = 51$.

To prove the theorem, we show that if a complex number z is such that $|z| > R_1$, then

$$|a_{n-1}z^{n-1} + \cdots a_1 z + a_0| < |z^n| \qquad (6.7)$$

This implies that z cannnot be a zero of $P_1(z)$, because zeros satisfy the equation

$$a_{n-1}z^{n-1} + \cdots a_1 z + a_0 = -z^n$$

which leads to

$$|a_{n-1}z^{n-1} + \cdots a_1 z + a_0| = |z^n|$$

We assume then that $|z| > R_1$, and obtain

$$
\begin{aligned}
|a_{n-1}z^{n-1} + \cdots a_1 z + a_0| &\leq |a_{n-1}||z|^{n-1} + \cdots |a_1||z| + |a_0| \\
&\leq (R_1 - 1)|z|^{n-1} + \cdots + (R_1 - 1)|z| + R_1 \\
&= (R_1 - 1)\frac{|z|^n - 1}{|z| - 1} + 1 \\
&< (|z| - 1)\frac{|z|^n - 1}{|z| - 1} + 1 = |z|^n
\end{aligned}
$$

The next result gives a way of forming a bounding radius that often is more accurate than R_1.

THEOREM 6.2: *If at least one of the coefficients a_i of the monic polynomial $P_1(z)$ is nonzero, the polynomial*

$$P_\rho(z) = z^n - |a_{n-1}|z^{n-1} - \cdots - |a_1|z - |a_0|$$

has exactly one positive zero ρ, which may be taken as a bounding radius for $P_1(z)$.

For our example polynomial, $z^4 - 10z^3 + 35z^2 - 50z + 24$, the polynomial $P_\rho(z)$ equals $z^4 - 10z^3 - 35z^2 - 50z - 24$, and its positive zero is 13.00~, so we get a smaller bounding radius than previously, though we need more computation to obtain it. To save computational effort, we can approximate the positive root of $P_\rho(z)$ by choosing some convenient small positive integer r, and then forming $P_\rho(r^n)$ for $n = 1, 2, \ldots$, stopping as soon as we have obtained $P_\rho(r^n) > 0$. The last value of r^n formed exceeds ρ, so we can use this last value as our bounding radius. With our example, if we choose $r = 2$, we find $P_\rho(2^3) < 0$ and $P_\rho(2^4) > 0$, so we get a bounding radius of 16 this way.

To prove the theorem, set z equal to the real number x, and write the polynomial $P_\rho(x)$ as

$$P_\rho(x) = x^n\left(1 - \frac{|a_{n-1}|}{x} - \cdots - \frac{|a_1|}{x^{n-1}} - \frac{|a_0|}{x^n}\right) \tag{6.8}$$

The factor x^n has no positive zeros, and the other factor has a positive derivative for x greater than zero, approaches $-\infty$ as x approaches zero from the right, and approaches 1 as x approaches ∞; so, this factor has exactly one positive zero ρ. If z is a complex number with $|z| > \rho$, then $P_\rho(|z|)$ is positive, and this implies

$$|z|^n > |a_{n-1}||z|^{n-1} + \cdots |a_1||z| + |a_0| \ge |a_{n-1}z^{n-1} + \cdots a_1z + a_0|$$

and again we have inequality 6.7. Therefore, z cannot be a zero of $P_1(z)$.

Sometimes the following easily calculated upper bound for ρ is useful.

COROLLARY: *The positive zero ρ satisfies the inequality*

$$\rho \le \max_{j=1}^n (n|a_{n-j}|)^{1/j}$$

We have

$$\rho^n = \sum_{i=0}^{n-1} |a_i|\rho^i$$

After we divide this equation by ρ^n we get

$$1 = \sum_{i=0}^{n-1} \frac{|a_i|}{\rho^{n-i}} = \sum_{j=1}^n \frac{|a_{n-j}|}{\rho^j}$$

Suppose the largest term in the second sum above occurs when the j index is j_0. We have then

$$1 \le n \cdot \frac{|a_{n-j_0}|}{\rho^{j_0}}$$

which leads to

$$\rho^{j_0} \le n|a_{n-j_0}|$$

$$\rho \le (n|a_{n-j_0}|)^{1/j_0}$$

Whatever the index j_0 is, ρ is bounded by $\max_{j=1}^n (n|a_{n-j}|)^{1/j}$. With our example polynomial, we find $\rho \le \max \{4 \cdot 10, (4 \cdot 35)^{\frac{1}{2}}, (4 \cdot 50)^{\frac{1}{3}}, (4 \cdot 24)^{\frac{1}{4}}\} = 40$.

6.3. NUMERICAL METHODS OF FINDING POLYNOMIAL ZEROS

There are many methods for finding approximations to the zeros of a polynomial, and we describe one efficient method for complex polynomials and one for real polynomials.

We assume that we want to find all the zeros of some complex polynomial $P(z)$. In this case an analytic version of Newton's method may be used. The Taylor series expansion for an analytic function $f(z)$ at a series expansion point w_k has the leading terms

$$f(z) = f(w_k) + f'(w_k)(z - w_k) + \cdots$$

If we follow the usual reasoning of Newton's method and approximate $f(z)$ by using just the first two terms of its series expansion, then setting $f(z)$ equal to zero and solving for z, calling our solution w_{k+1}, we obtain the complex iteration equation

$$w_{k+1} = w_k - \frac{f(w_k)}{f'(w_k)} \tag{6.9}$$

This iteration equation can be used to find zeros of a complex polynomial $P(z)$ if the progress of the iteration is closely monitored. A convenient point w_0 is chosen within the zero bounding circle $|z| \le R$ (see Section 6.2), and as each iterate w_k is found, some indicator of progress toward a zero is also calculated. The indicator could be $|P(w_k)|$ or the more easily calculated sum of the absolute values of the real and imaginary parts of $P(w_k)$. If the indicator decreases in value, this is taken as evidence that the new iterate w_{k+1} is closer to a zero than w_k. If the indicator does not decrease in going from w_k to w_{k+1}, then one can try moving w_{k+1} back half the distance toward w_k one or more times to see if this results in an improved indicator. If not, the iteration is abandoned and a new initial point w_0 is chosen.

For the complex number $w_k = x_k + iy_k$, define mid w_k to be mid $x_k + i$(mid y_k), and define $w_k \doteq w_{k+1}$ to mean $x_k \doteq x_{k+1}$ and $y_k \doteq y_{k+1}$ After each iteration cycle we check if $w_{k+1} \doteq w_k$, and if this relation is not true, then w_{k+1} is replaced by mid w_{k+1} in preparation for the next iteration cycle. After we finally obtain $w_{k+1} \doteq w_k$, then mid w_k can serve as the final zero approximation w_Z. If the degree of the problem polynomial is greater than 1, the factor $z - w_Z$ is divided into the problem polynomial to get a problem polynomial of lower degree, and the whole process is repeated with the new problem polynomial.

Newton's method in its many forms works well when the iteration point is *close enough* to the target point, but various types of oscillation may appear when this is not the case. When finding zeros of polynomials, it may happen that an iterate w_k lands on a line L in the complex plane that separates the polynomial zeros symmetrically. That is, for each polynomial zero on one side of the line L there is another zero that is its mirror image in the line. In this case the successive iterates remain on L, and thus if there are no polynomial zeros on L, the iterates cycle endlessly on the line. If the polynomial is real, the real line is a line of symmetry for the polynomial, because the complex zeros of a real polynomial occur in conjugate pairs. In this case note that if we take w_0 as real, then successive iterates are real also. The indicator system would be needed to restart the iteration at a new initial point w_0.

Suppose now that our problem polynomial $P(z)$ is real. A real polynomial of even degree can be written as a product of real quadratic polynomials, because in the factorization 6.2 each pair of real zeros r_1, r_2 yields the real quadratic factor

$$(z - r_1)(z - r_2) = z^2 - (r_1 + r_2)z + r_1 r_2$$

and a conjugate pair of zeros $x_1 + iy_1$, $x_1 - iy_1$ yields the real factor

$$(z - x_1 - iy_1)(z - x_1 + iy_1) = [(z - x_1) - iy_1][(z - x_1) + iy_1]$$

$$= (z - x_1)^2 - (iy_1)^2 = z^2 - 2x_1 z + x_1^2 + y_1^2$$

A real polynomial of odd degree can be written as a product of real quadratic polynomials and one real linear polynomial.

The Bairstow method for real polynomials, like the preceding method for complex polynomials, is a variation on Newton's method. The method finds quadratic real factors, and in this way finds zeros, a pair at a time. Suppose $B(z) = z^2 + b_1 z + b_0$ is an approximation to some quadratic factor of the real polynomial $P(z)$ of degree n. When we divide $P(z)$ by $B(z)$, we obtain the quotient $Q(z)$ and a linear remainder $c_1 z + c_0$:

$$P(z) = B(z)Q(z) + c_1 z + c_0 \tag{6.10}$$

The coefficients c_1 and c_0 are functions of b_1 and b_0, and we want to choose the variables b_1 and b_0 so both $c_1(b_1, b_0)$ and $c_0(b_1, b_0)$ are zero. This is an instance of the problem we considered in Chapter 5, solving the vector equation $\mathbf{f}(\mathbf{x}) = \mathbf{0}$, and the Newton's iteration equation used there was

$$\mathbf{x}^{(k+1)} = \mathbf{x}^{(k)} - [J(\mathbf{x}^{(k)})]^{-1}\mathbf{f}(\mathbf{x}^{(k)})$$

When we reinterpret this equation for our problem, we get the iteration equation

$$\begin{bmatrix} b_1^{(k+1)} \\ b_0^{(k+1)} \end{bmatrix} = \begin{bmatrix} b_1^{(k)} \\ b_0^{(k)} \end{bmatrix} - \begin{bmatrix} \dfrac{\partial c_1}{\partial b_1} & \dfrac{\partial c_1}{\partial b_0} \\ \dfrac{\partial c_0}{\partial b_1} & \dfrac{\partial c_0}{\partial b_0} \end{bmatrix}^{-1} \begin{bmatrix} c_1 \\ c_0 \end{bmatrix}$$

To get expressions for the matrix partial derivatives, we differentiate Equation 6.10 with respect to b_1 first and then with respect to b_0, obtaining the equations

$$0 = zQ(z) + B(z)\frac{\partial Q(z)}{\partial b_1} + \frac{\partial c_1}{\partial b_1}z + \frac{\partial c_0}{\partial b_1} \tag{6.11}$$

$$0 = Q(z) + B(z)\frac{\partial Q(z)}{\partial b_0} + \frac{\partial c_1}{\partial b_0}z + \frac{\partial c_0}{\partial b_0} \tag{6.12}$$

Suppose we write either of these equations in the form

$$0 = B(z)S(z) + \alpha z + \beta$$

When we equate the right-hand coefficient of z^k to zero, starting with k large on down to $k = 2$, we can conclude that all coefficients of $S(z)$ are zero, and this in turn implies that α and β are zero. To carry this plan out, we divide $Q(z)$ by $B(z)$ to obtain the quotient $Q_1(z)$ and the remainder $d_1z + d_0$:

$$Q(z) = B(z)Q_1(z) + d_1z + d_0$$

If we substitute this result in Equation 6.12, we obtain the equation

$$0 = B(z)\left[Q_1(z) + \frac{\partial Q(z)}{\partial b_0}\right] + \left(d_1 + \frac{\partial c_1}{\partial b_0}\right)z + \left(d_0 + \frac{\partial c_0}{\partial b_0}\right)$$

which implies

$$\frac{\partial c_1}{\partial b_0} = -d_1$$

$$\frac{\partial c_0}{\partial b_0} = -d_0$$

Making a similar substitution in Equation 6.11, and replacing the term d_1z^2 by $d_1B(z) - d_1b_1z - d_1b_0$, we get

$$0 = B(z)\left[zQ_1(z) + \frac{\partial Q(z)}{\partial b_1} + d_1\right] + \left(d_0 - d_1b_1 + \frac{\partial c_1}{\partial b_1}\right)z + \left(-d_1b_0 + \frac{\partial c_0}{\partial b_1}\right)$$

which implies

$$\frac{\partial c_1}{\partial b_1} = d_1b_1 - d_0$$

$$\frac{\partial c_0}{\partial b_1} = d_1b_0$$

We have then

$$\begin{bmatrix} \dfrac{\partial c_1}{\partial b_1} & \dfrac{\partial c_1}{\partial b_0} \\[2ex] \dfrac{\partial c_0}{\partial b_1} & \dfrac{\partial c_0}{\partial b_0} \end{bmatrix} = \begin{bmatrix} d_1 b_1 - d_0 & -d_1 \\[1ex] d_1 b_0 & -d_0 \end{bmatrix}$$

with a determinant $D = d_1^2 b_0 - (d_1 b_1 - d_0) d_0$, and the matrix inverse is

$$\begin{bmatrix} \dfrac{\partial c_1}{\partial b_1} & \dfrac{\partial c_1}{\partial b_0} \\[2ex] \dfrac{\partial c_0}{\partial b_1} & \dfrac{\partial c_0}{\partial b_0} \end{bmatrix}^{-1} = \frac{1}{D}\begin{bmatrix} -d_0 & d_1 \\[1ex] -d_1 b_0 & d_1 b_1 - d_0 \end{bmatrix}$$

The Bairstow method consists of choosing some initial quadratic $z^2 + b_1^{(0)} z + b_0^{(0)}$ and then performing the following cycle: The problem polynomial $P(z)$ is divided by $B^{(k)}(z) = z^2 + b_1^{(k)} z + b_0^{(k)}$ to obtain the quotient polynomial $Q(z)$ and the remainder $c_1 z + c_0$. Then $Q(z)$ is divided by $B^{(k)}(z)$ to obtain a quotient and a remainder $d_1 z + d_0$. The quadratic coefficients $b_1^{(k+1)}$, $b_0^{(k+1)}$ are formed from $b_1^{(k)}$, $b_0^{(k)}$ by adding the amounts

$$\Delta b_1 = \frac{c_1 d_0 - c_0 d_1}{D} \qquad \Delta b_0 = \frac{c_1 d_1 b_0 - c_0(d_1 b_1 - d_0)}{D}$$

If we find $b_1^{(k+1)} \neq b_1^{(k)}$ or find $b_0^{(k+1)} \neq b_0^{(k)}$, then, in preparation for the next cycle, $b_1^{(k+1)}$ and $b_0^{(k+1)}$ are replaced by mid $b_1^{(k+1)}$ and mid $b_0^{(k+1)}$, respectively. Otherwise the iteration ends and the final quadratic approximation $z^2 + b_1 z + b_0$ uses the mid values of the k-th iterates, from which we obtain two zeros of $P(z)$ by using the quadratic formula $-\frac{b_1}{2} \pm \sqrt{(\frac{b_1}{2})^2 - b_0}$. The quotient $Q(z)$ becomes the new problem polynomial, and another cycle of quadratic approximations begins, unless of course the degree of our problem polynomial has become two or less, in which case the remaining zeros are found without difficulty.

As with the Newton's iteration method for complex polynomials, the Bairstow iteration must be monitored for steady progress toward a quadratic factor of $P(z)$. The value of $|c_1| + |c_0|$ can be used as an indicator. If this indicator does not decrease with an iteration, then Δb_1 and Δb_0 can be changed to a fraction of their previous values, and this test repeated. If a reduced indicator is not obtained after one or more such rescue attempts, the whole procedure is restarted by choosing new starting values for $b_1^{(0)}$ and $b_0^{(0)}$.

6.4. BOUNDING THE ERROR OF POLYNOMIAL ZERO APPROXIMATIONS

Suppose for some monic real or monic complex polynomial $P(z)$ of degree n, we have found distinct zero approximations $w_1, w_2, \ldots, w_q$ with respective apparent multiplicities $n_1, n_2, \ldots, n_q$ summing to n. These approximations are the actual zeros of a certain polynomial $Q(z)$ of degree n, which has the factorization

$$Q(z) = (z - w_1)^{n_1}(z - w_2)^{n_2} \cdots (z - w_q)^{n_q}$$

The rational function $P(z)/Q(z)$ has the partial fraction expansion

$$\frac{P(z)}{Q(z)} = 1 + \sum_{i=1}^{q} \sum_{j=1}^{n_i} \frac{h_{ij}}{(z - w_i)^j} \tag{6.13}$$

with complex coefficients h_{ij}. If we set z equal to any zero z_p of $P(z)$, we get the equation

$$-1 = \sum_{i=1}^{q} \sum_{j=1}^{n_i} \frac{h_{ij}}{(z_P - w_i)^j}$$

Taking absolute values, we obtain the relations

$$1 = \left| \sum_{i=1}^{q} \sum_{j=1}^{n_i} \frac{h_{ij}}{(z_P - w_i)^j} \right| \leq \sum_{i=1}^{q} \sum_{j=1}^{n_i} \frac{|h_{ij}|}{|z_P - w_i|^j}$$

Suppose the double sum term $\frac{|h_{ij}|}{|z_P - w_i|^j}$ is largest for $i = i_0$ and $j = j_0$. Then we have

$$1 \leq n \frac{|h_{i0,j0}|}{|z_P - w_{i0}|^{j_0}}$$

$$|z_P - w_{i0}|^{j_0} \leq n|h_{i0,j0}|$$

$$|z_P - w_{i0}| \leq (n|h_{i0,j0}|)^{1/j_0}$$

Now suppose for each zero approximation w_i we compute the quantity

$$\epsilon_i = \max_{j=1}^{n_i} (n|h_{ij}|)^{1/j} \tag{6.14}$$

Then it is certain that every zero of $P(z)$ lies in one of the disks

$$|z - w_i| \leq \epsilon_i \qquad i = 1, 2, \ldots, q \tag{6.15}$$

It is also possible to conclude that if these disks do not intersect each other, then the disk associated with the approximation w_i of apparent multiplicity n_i contains exactly n_i zeros of $P(z)$. Moreover, if several of these disks overlap, and the sum of the associated multiplicities of these disks is m, then the composite figure contains exactly m zeros of $P(z)$. To see this, consider the polynomial

$$P_\lambda(z) = \lambda P(z) + (1 - \lambda)Q(z)$$

where the real parameter λ varies in the interval $[0, 1]$. The polynomial $P_\lambda(z)$ equals $Q(z)$ when $\lambda = 0$ and equals $P(z)$ when $\lambda = 1$. We have

$$\frac{P_\lambda(z)}{Q(z)} = 1 + \sum_{i=1}^{q} \sum_{j=1}^{n_i} \frac{\lambda h_{ij}}{(z - w_i)^j} \tag{6.16}$$

and according to our previous analysis, each zero of $P_\lambda(z)$ lies in one of the disks

$$|z - w_i| \leq \epsilon_i(\lambda) = \max_{j=1}^{n_i} (n|\lambda h_{ij}|)^{1/j} \ i = 1, 2, \ldots, q$$

The radius $\epsilon_i(\lambda)$ varies with λ, being zero when $\lambda = 0$ and equal to the radius of the $P(z)$ disk when $\lambda = 1$. When λ is 0, the point disk at w_i has a zero of multiplicity n_i of $P_0(z) = Q(z)$. As λ varies in $[0, 1]$, the various zeros of $P_\lambda(z)$ are continuous functions of λ. Because the number of zeros in all the disks is always n, the number of zeros of $P_\lambda(z)$ in the disk with center w_i must remain equal to n_i as long as this disk does not intersect any of the other disks. Otherwise there would be a discontinuous jump of at least one zero of $P_\lambda(z)$. Similarly, when λ is of a size that a certain number of the $P_\lambda(z)$ disks overlap to form a composite structure, the total number of zeros of $P_\lambda(z)$ in the overlapping disks must equal the sum of the associated multiplicities.

The general plan for computing error bounds is as follows: Suppose we have a set of $P(z)$ zero approximations w_i, computed by a method described in the preceding section or by some other method. If these approximations are not already exact, they are made exact, that is, the approximations w_i are replaced by mid w_i before beginning the error computation. These zero approximations have apparent multiplicities assigned if the $N_i(z)$ decomposition of $P(z)$ was employed. If this decomposition was not used, then the approximations need to be compared with each other, with m identical approximations combined into a single approximation of apparent multiplicity m. After the various quantities h_{ij} are computed, by a method described later in this section, the radius ϵ_i is obtained for each approximation $w_i = x_i + iy_j$, and this radius is used to assign appropriate wids to w_i, changing it to $\hat{w}_i = (x_i \ominus \epsilon_i) + i(y_i \ominus \epsilon_i)$. The next step is to compare the approximations $\hat{w}_i$ with each other to make certain that no two overlap, that is, to make certain that $\hat{w}_i \neq \hat{w}_j$ for $i \neq j$. If some of them overlap, the original approximations w_i need to be combined corresponding to the overlap, choosing arbitrarily one approximation w_i to represent the set of overlapping approximations and increasing apparent multiplicities appropriately. Then the error bound computation is repeated. When finally there is no overlap, an approximation w_i of apparent multiplicity m is a correctly ranged approximation to m zeros of $P(z)$, though these m zeros need not all be equal.

Suppose we are trying to compute zero approximations to a certain number of correct fixed-point decimals, and we have a zero approximation w_i of apparent multiplicity m. In order that ϵ_i be no greater than 10^{-k}, according to Equation 6.14 we must have $|nh_{i,m}|$ no larger than 10^{-km}. We see then, that to achieve k correct decimals, we need to carry around km decimals in our computations for these approximations. Multiple zeros require higher precision computation than simple zeros, with the precision needed being proportional to the multiplicity.

We consider next the method of computing the needed quantities $h_{i,j}$. This essentially is a Taylor series computation. If we multiply Equation 6.13 by $(z - w_i)^{n_i}$, we obtain

$$\frac{P(z)}{\displaystyle\prod_{\substack{j=1 \\ j \neq i}}^{n} (z - w_j)^{n_j}} = h_{i,\,n_i} + h_{i,\,n_{i-1}}(z - w_i) + \cdots h_{i,1}(z - w_i)^{n_i - 1} + \cdots$$

On the right side of the equals sign we have collected all the terms that contain powers of $(z - w_i)$ with an exponent less than n_i. The other terms on the right side can themselves be expanded in a Taylor series about the expansion point w_i, but the factor $(z - w_i)^{n_i}$ that they have prevents any terms in $(z - w_i)^k$ appearing with an exponent k less than n_i. Thus the terms in $(z - w_i)$ shown above are the correct leading Taylor series terms for the function on the left. Therefore we will obtain the needed coefficients h_{ij} for a particular i, if we find for the function on the left above, its leading Taylor series terms about the expansion point w_i. We need the two expansions

$$P(z) = b_0 + b_1(z - w_i) + b_2(z - w_i)^2 + \cdots + (z - w_i)^{n_i - 1} + \cdots$$

$$\frac{Q(z)}{(z - w_i)^{n_i}} = \prod_{\substack{j=1 \\ j \neq i}}^{n} (z - w_j)^{n_j} = c_0 + c_1(z - w_i) + c_2(z - w_i)^2 + \cdots + (z - w_i)^{n_i - 1} + \cdots$$

and then we can form the needed terms by doing a series division, using the relation 4.7 of Chapter 4. For this division we need $c_0 \neq 0$, which we have if our precision is high enough because $w_j \neq w_i$ for $j \neq i$.

6.5. A METHOD TO OBTAIN CORRECTLY RANGED POLYNOMIAL ZEROS

When the zeros of a polynomial are sought numerically, simple zeros are easier to locate accurately than multiple zeros. For instance, according to Section 5.4, if we find a real zero of a polynomial by Newton's method, for a simple zero the convergence is quadratic, while for a multiple zero the convergence is linear and is slower the greater the multiplicity is. In Section 6.3 two methods were described for locating complex zeros of a polynomial, and because both methods are adaptations of Newton's method, we can expect similar slow convergence when trying to locate a multiple zero. At a simple zero the polynomial has a nonzero rate of change, while at a multiple zero the polynomial's derivative is zero, and the polynomial changes at a more leisurely pace making the determination of the exact position of the zero more difficult. To avoid the troublesome problem of finding multiple zeros, our demonstration programs use the Euclidean procedure described in Section 6.1 to replace the problem polynomial $P(z)$ with the polynomials $N_1(z)$, $N_2(z)$, . . . , $N_q(z)$ having only simple zeros.

When $P(z)$ is a rational or a complex rational polynomial, the polynomials $N_i(z)$ also make it easier to compute the error of the zero approximations. After the zero approximations of a polynomial $N_i(z)$ are assembled, we can compute their error bounds as simple zeros of $N_i(z)$, rather than as possibly multiple zeros of the original problem polynomial $P(z)$. If the $N_i(z)$ zeros are not found to the desired number of correct decimal places, we need only increase the precision of computation and repeat the zero finding and error bounding procedure on $N_i(z)$ alone.

For a real or complex polynomial $P(z)$, the Euclidean procedure can be used to obtain polynomials $N_i(z)$, but here there is ambiguity whether the computed polynomials $N_i(z)$ are correct. Consider the range arithmetic division of the Euclidean procedure, where a polynomial $R_j(z)$ is divided by a polynomial $R_{j+1}(z)$ of degree k. The remainder polynomial has the general form

$$d_{k-1}z^{k-1} + d_{k-2}z^{k-2} + \cdots d_1z + d_0$$

Suppose we find that a certain number of leading coefficients approximate zero, that is, $0 \doteq d_{k-1} \doteq d_{k-2} \doteq \ldots \doteq d_s$. We take this result to mean that $0 = d_{k-1} = d_{k-2} = \ldots = d_s$, and if d_s is d_0, the Euclidean procedure has terminated, otherwise the next polynomial $R_{j+2}(z)$ is of degree $s - 1$. But perhaps one or more of the coefficients presumed to be zero are actually not zero. In that case the precision of computation should have been increased and the Euclidean procedure redone. Thus with range arithmetic, at any given precision of computation, the Euclidean procedure yields polynomials $N_i(z)$, but we are uncertain whether they are correct. When the zero approximations for all polynomials $N_i(z)$ are assembled, and we are ready to bound their error, we must use the original polynomial $P(z)$ in the error bounding process. If it turns out that not enough correct decimals have been determined, we must increase the precision of computation appropriately and repeat all steps of our procedure, and that includes a new determination of the polynomials $N_i(z)$. If it turns out that we are d decimal places short of our correct decimal place goal, and $N_q(z)$ is the last polynomial of the $N_i(z)$ set, indicating the probable presence of a zero of highest multiplicity q, then as indicated in Section 6.4 precision should be increased by about qd decimal places.

6.6. THE ELEMENTARY ANALYTIC FUNCTIONS

To discuss computational problems with analytic functions, it is convenient to define a certain class of functions, the *elementary* analytic functions, analogous to the elementary real functions of previous chapters. Like elementary real functions, the elementary analytic functions are frequently encountered functions for which it is easy to form power series and to compute derivatives. Just as for an elementary real function, an elementary analytic functions may use a finite number of certain operations. Elementary real functions may employ two argument max or min evaluations, but these operations cannot be extended sensibly into the complex domain, so are not included in the corresponding list of allowed operations. All the single argument evaluation operations possible for real elementary functions, however, have complex extensions.

DEFINITION 6.1: A function of a finite number of complex variables z_1, z_2, z_3, . . . is an *elementary* analytic function if it can be expressed in terms of variables and constants using a finite number of the binary operations of addition, subtraction, multiplication, division, and exponentiation and the operation of single argument function evaluation with sin, cos, tan, $\sin^{-1}$, $\cos^{-1}$, $\tan^{-1}$, exp, or log.

The complete sequence of operations needed to compute an elementary analytic function is to be presumed known, just as it was for an elementary real function.

The series relations developed in Chapter 4 for computing function values and derivatives of real elementary functions also apply to the elementary analytic functions, with the exception of course of the max and min relations. In particular we have the key relation

$$f(z) = f(z_0) + \frac{f'(z_0)}{1!}(z - z_0) + \cdots + \frac{f^{(n-1)}(z_0)}{(n-1)!}(z - z_0)^{n-1}$$
$$+ \frac{f^{(n)}(z_0) \pm w}{n!}(z - z_0)^n$$

Here $f^{(n)}(z_0) \pm w$ denotes a real part interval together with an imaginary part interval. There are some new issues concerning the evaluation of the complex standard functions that need to be addressed, and we take these up next.

The inverse of the exponential function $e^z = e^x(\cos x + i \sin x)$ is $\log z$, and it is well known that it is not possible to define $\log z$ in the complex plane without introducing a so-called cut line on which the function is discontinuous. This cut line is usually taken as the negative real axis, and the definition of log z is then $\ln |z| + i\theta(z)$, where $\theta(z)$ gives the *principal argument* of z, a value between π and $-\pi$. As z approaches a point $-x_0$ on the negative real axis from above, log z approaches $\ln x_0 + i\pi$, and as z approaches this point from below, log z approaches $\ln x_0 - i\pi$, so log z is discontinuous on the cut line. If we compute log z with range arithmetic, we must be resigned to having our function undefined on the negative real axis (see Section 3.5). Accordingly in this text log z is taken as undefined on the negative real axis.

Let a be any *real* constant but not an integer. The exponentiation operation z^a may be taken as

$$z^a = e^{a \log z} \tag{6.17}$$

and is defined throughout the complex z plane except at the origin and on the negative real axis. For irrational a, the real function x^a is normally taken as

$$x^a = \begin{cases} 0^a & \text{if } x = 0 \\ \text{undefined} & \text{if } x < 0 \\ e^{a \ln x} & \text{otherwise} \end{cases}$$

If we take line 6.17 as our definition of z^a, then our function is undefined for negative reals and $z = 0$ and is in agreement with the real function x^a for real z, except when $a > 0$, for then x^a is defined at the origin and z^a is not. When extending a real function to allow complex arguments, a reasonable principle to adhere to is that if the real function is defined at some argument, its extension should not become undefined or change its value at this argument. Accordingly, we take z^a to be

$$z^a = (x + iy)^a = \begin{cases} 0^a & \text{if } x = 0 \text{ and } y = 0 \\ \text{undefined} & \text{if } x < 0 \text{ and } y = 0 \\ e^{a \log z} & \text{otherwise} \end{cases} \tag{6.18}$$

Requiring z^a to equal zero at $z = 0$ does not introduce a discontinuity, and the limit of z^a is zero as z approaches the origin along any ray on which z^a is defined. In particular, for the important case $a = \frac{1}{2}$ we have

$$\sqrt{z} = z^{1/2} = (x + iy)^{1/2} = \begin{cases} 0 & \text{if } x = 0 \text{ and } y = 0 \\ \text{undefined} & \text{if } x < 0 \text{ and } y = 0 \\ e^{(1/2) \log z} & \text{otherwise} \end{cases} \tag{6.19}$$

If a is a rational number, and p/q is in reduced form with the integer q positive, the usual interpretation of x^a is

$$x^{p/q} = (\sqrt[q]{x})^p = \sqrt[q]{x^p}$$

When q is even, Equation 6.18 leads to a definition of $z^{1/q}$ that is in accordance with $x^{1/q} = \sqrt[q]{x}$, namely,

$$q \text{ even: } \sqrt[q]{z} = z^{1/q} = (x + iy)^{1/q} = \begin{cases} 0 & \text{if } x = 0 \text{ and } y = 0 \\ \text{undefined} & \text{if } x < 0 \text{ and } y = 0 \\ e^{(1/q) \log z} & \text{otherwise} \end{cases}$$

(6.20)

When q is odd we cannot use (6.20) because then $z^{1/q}$ would be undefined for negative real z, whereas $x^{1/q}$ is defined for negative real x. A possible definition of $z^{1/q}$ when q is odd is

$$q \text{ odd: } \sqrt[q]{z} = z^{1/q} = (x + iy)^{1/q} = \begin{cases} 0 & \text{if } x = 0 \text{ and } y = 0 \\ \text{undefined} & \text{if } x = 0 \text{ and } y \neq 0 \\ e^{(1/q) \log z} & \text{if } x > 0 \\ -e^{(1/q) \log -z} & \text{if } x < 0 \end{cases}$$

(6.21)

On the right half of the complex plane we use 6.20, but on the left half we use a different q-th root, with negative real part, so that $z^{1/q}$ have the same value for negative real z as does $x^{1/q}$. We have $(-e^{(1/q) \log -z})^q = (-1)^q e^{\log -z} = (-1)^q(-z) = (-1)^{q+1}z = z$, so the assigned value is a q-th root. On the imaginary axis, except at the origin, we take our function as undefined, because it would be discontinuous on this line whatever value we assigned.

If $c = a + ib$ is a *complex* number, we define z^c as

$$z^c = (x + iy)^{a+ib} = \begin{cases} 0^a & \text{if } x = 0 \text{ and } y = 0 \\ \text{undefined} & \text{if } x < 0 \text{ and } y = 0 \\ e^{c \log z} & \text{otherwise} \end{cases}$$

This definition extends the pattern we used for real exponents.

The inverse functions $\sin^{-1} z$, $\cos^{-1} z$, and $\tan^{-1} z$ can be defined by the equations

$$\sin^{-1} z = -i \log[iz + \sqrt{1 - z^2}]$$

$$\cos^{-1} z = \frac{\pi}{2} - \sin^{-1} z$$

$$\tan^{-1} z = \frac{i}{2} \log \frac{i + z}{i - z}$$

Using the interpretation 6.19 for the complex square-root, these equations imply that $\sin^{-1} z$ and $\cos^{-1} z$ are undefined when $1 - z^2$ is negative real, which occurs when z is real and greater in absolute value than 1. These are the only points where these two functions are undefined, because in the equation for $\sin^{-1} z$, the argument of its log function never takes on a negative real value. On the other hand, for $\tan^{-1} z$, the argument of its log function is negative real for $z = iy$ with $|y| > 1$, and the log function is not defined when y equals 1 or -1. So $\tan^{-1} z$ is not defined on the imaginary axis except for the open interval between -1 and $+1$.

6.7. FINDING A ZERO OF AN ELEMENTARY ANALYTIC FUNCTION

Suppose for an elementary analytic function $f(z)$, we want to find all the zeros in some rectangle R in the complex plane, defined by the two inequalities $a \leq x \leq b$ and $c \leq y \leq d$. The function $f(z)$ is presumed defined throughout R. This problem may be considered equivalent to the problem of finding where two real functions of two real variables are zero, in a region specified by bounding each of the two real variables in a finite interval. We have $f(z) = f(x + iy) = u(x, y) + iv(x, y)$, and the two real functions are $u(x, y)$ and $v(x, y)$. When we considered such real variable problems in Chapter 5 to obtain a solvable problem, it was necessary to require that there be no zeros on the boundary of the search domain. For the same reason $f(z)$ is now required to be nonzero on the boundary of R. Also if an elementary analytic function is zero on some curve inside R, then it must be identically zero. So if we know the function $f(z)$ is nonzero on the boundary of R, we know that its zeros in R are isolated zeros. This makes our solvable problem for elementary analytic functions simpler than the real variable variety.

SOLVABLE PROBLEM 6.4: For any elementary analytic function $f(z)$ defined in a rectangle R on the boundary of which $f(z)$ is nonzero, bound all the zeros of f in R by giving to k decimals the real and imaginary parts of

1. Points identified as simple zeros, where $f'(z)$ is nonzero, or
2. Points identified as zeros of apparent multiplicity m greater than one.

If a real number r is displayed to k correct decimals, the number r may lie anywhere in the corresponding tilde interval. Similarly, in the complex plane a complex number $z = x + iy$ displayed to k decimals may lie anywhere in a tilde rectangle centered at the z displayed value, with sides parallel to the real and imaginary axes. If a zero of apparent multiplicity m is displayed to k decimals, there are exactly m zeros within the tilde rectangle if we add the multiplicities of the various individual zeros. For instance, If m is 3, then within the tilde rectangle there could be one zero of multiplicity 3, there could be three simple zeros, or there could be a simple zero and a zero of multiplicity 2.

If the analytic function $f(z)$ is defined inside and on the rectangle R and $f(z)$ is nonzero on R, then the number of zeros inside R can be determined by making use of the *argument principle*, namely that the number of zeros in R, counting multiplicities, equals the integral

$$\frac{1}{2\pi i} \int_R \frac{f'(z)}{f(z)} \, dz \qquad (6.22)$$

where the integration around R is done in the counterclockwise direction. If $f(z)$ equals $u(x, y) + iv(x, y)$, we may view the complex function f as equivalent to the mapping $\mathbf{f}$ defined by $u(x, y)$ and $v(x, y)$ over the rectangle R in the *real cartesian* plane, instead of the complex plane. The integral 6.22 equals the topological degree of the function $\mathbf{f}$. To see this, examine any of the four parts of the integral over sides of R. For instance, the contribution to the integral 6.22 by the bottom edge of R is

$$\frac{1}{2\pi i} \int_a^b \frac{\dfrac{\partial u}{\partial x} + i \dfrac{\partial v}{\partial x}}{u + iv} \, dx = \frac{1}{2\pi i} \int_a^b \frac{u \dfrac{\partial u}{\partial x} + v \dfrac{\partial v}{\partial x} + i\left(-v \dfrac{\partial u}{\partial x} + u \dfrac{\partial v}{\partial x}\right)}{u^2 + v^2} \, dx$$

$$= \frac{1}{2\pi i} \cdot \frac{1}{2} \log(u^2 + v^2)\Big|_a^b + \frac{1}{2\pi} \int_a^b \frac{\det \begin{bmatrix} u & v \\ \dfrac{\partial u}{\partial x} & \dfrac{\partial v}{\partial x} \end{bmatrix}}{u^2 + v^2} \, dx$$

The log term of the last equation above may be ignored because it is cancelled by similar terms from the other sides of R. The other term is recognizable as the appropriate part of $td(\mathbf{f}, R)$ over the bottom edge of R.

The argument principle implies that for an elementary analytic function $f(z)$, the topological degree of the $\mathbf{f}$ function is always positive, equaling the number of zeros in R. This is easily seen for the case where $f(z)$ has only simple zeros, for then $td(\mathbf{f}, R)$ equals the number of $\mathbf{f}$ zeros in R with a positive Jacobian minus the number with a negative Jacobian. Making use of the Cauchy-Riemann equations $\frac{\partial u}{\partial x} = \frac{\partial v}{\partial y}$ and $\frac{\partial u}{\partial y} = \frac{-\partial v}{\partial x}$, we obtain

$$\frac{\partial(u, v)}{\partial(x, y)} = \det \begin{bmatrix} \dfrac{\partial u}{\partial x} & \dfrac{\partial u}{\partial y} \\ \dfrac{\partial v}{\partial x} & \dfrac{\partial v}{\partial y} \end{bmatrix} = \det \begin{bmatrix} \dfrac{\partial u}{\partial x} & \dfrac{\partial u}{\partial y} \\ -\dfrac{\partial u}{\partial y} & \dfrac{\partial u}{\partial x} \end{bmatrix} = \left(\dfrac{\partial u}{\partial x}\right)^2 + \left(\dfrac{\partial u}{\partial y}\right)^2 = |f(z)|^2$$

Because the Jacobian is positive at each simple zero, the topological degree equals the number of zeros in R.

If we always know the number of zeros of $f(z)$ inside a problem rectangle R, this makes the problem of determining those zeros much easier. The search procedure can be done with a task queue holding problem rectangles to be investigated, the task queue initially holding just R. If R_Q is the first rectangle on the queue, let $n(R_Q)$ be the number of zeros inside R_Q obtained by computing $td(\mathbf{f}, R_Q)$. If $n(R_Q) = 0$, we can discard R_Q. If $n(R_Q) = 1$, we try to find the simple zero by a complex version of Newton's method, using the centerpoint of our rectangle as the initial trial point z_0. If there is some difficulty with the iteration, say the iterates z_i repeatedly fall outside R_Q, then we can try using a few other initial values for z_0, and if these fail too, we bisect R_Q in its longer dimension, find which subrectangle contains the zero by computing the topological degree for either subrectangle, discard the unneeded subrectangle, and resume our Newton's method attempt with a smaller rectangle. Eventually we determine the simple zero to the required number of decimal places.

If $n(R_Q) > 1$, we bisect R_Q by a division of its larger side or a division of either side if R is a square, obtaining two problem rectangles R_{Q_1} and R_{Q_2}. A computation of $n(R_{Q_1})$ also determines $n(R_{Q_2})$, because $n(R_Q) = n(R_{Q_1}) + n(R_{Q_2})$. However, if there is a multiple zero z_m inside R_Q with multiplicity equal to $n(R_Q)$, we obtain a series of subrectangles enclosing z_m, and for these subrectangles the zero count never is one, which is our criterion for using Newton's method. The size of the subrectangles surrounding z_m continues to diminish until one is obtained that falls within the tilde rectangle of its centerpoint if it were displayed to the requisite k decimal places. The centerpoint then is displayed but only as a zero of *apparent* multiplicity m, because, as far as we know, inside it there could be several separate zeros with multiplicities summing to m.

We have glossed over a potential difficulty in the procedure described, in that when we bisect R_Q, it may happen that a zero lies on the common side of the two subrectangles. In this case we cannot compute the topological degree for $f(z)$ over a subrectangle, and so cannot determine the zero count for either subrectangle. We must have a means of avoiding this trouble. Here knowing $n(R_Q)$ is helpful. Before we accept a bisection of R_Q, we can do an interval arithmetic computation of $f(z)$ over the segment that cuts R_Q. If the interval for the real part or the imaginary part of $f(z)$ does not contain the zero point, then the bisection is accepted. Otherwise, we divide the cut segment into two subsegments and try the interval computation again for each subsegment. This is another task queue computation, where we continue subdividing until the subsegment wid at the head of the queue passes below a present bound δ, at which point the process is halted. Of course if the task queue becomes empty before this happens, the bisection is accepted. In the case of bisection failure, a new division line parallel to but somewhat removed from the previous bisection line is chosen, and the cut segment check is repeated. The division line can be changed each time bisection fails, but when the number of lines tested exceeds $n(R_Q)$, this signals that either our precision of computation is too small, or the δ bound we are using is too large, and the entire bisection process can now be repeated from the beginning using a higher precision and a smaller δ. Eventually we must be successful in finding an acceptable division line.

6.8. THE GENERAL ZERO PROBLEM FOR ELEMENTARY ANALYTIC FUNCTIONS

The problem we consider here is finding where n elementary analytic functions of n variables are simultaneously zero. The equations that must be solved are

$$
\begin{aligned}
f_1(z_1, z_2, \ldots, z_n) &= 0 \\
f_2(z_1, z_2, \ldots, z_n) &= 0 \\
&\vdots \\
f_n(z_1, z_2, \ldots, z_n) &= 0
\end{aligned}
\tag{6.23}
$$

The region of interest is defined by restricting each variable z_i to a rectangle of its complex domain or, equivalently, restricting the real part and the imaginary part of each variable $z_i = x_i + iy_i$ to a finite interval:

$$
a_i \le x_i \le b_i \qquad \text{and} \qquad c_i \le y_i \le d_i, \qquad i = 1, 2, \ldots, n \tag{6.24}
$$

Again to obtain a solvable problem, it must be required that there be no zeros on the boundary of the search domain.

It is convenient to use vector notation once more, with $\mathbf{z}$ denoting a vector with components $z_1, z_2, \ldots, z_n$ and $\mathbf{f}(\mathbf{z})$ denoting the function with the components 6.23. The function $\mathbf{f}$ is counted elementary analytic if all its components are elementary analytic. For lack of any better term, we still call the domain 6.24 a box. The appropriate generalization of Problem 6.4 is

SOLVABLE PROBLEM 6.5: For any elementary analytic function $\mathbf{f}(\mathbf{z})$ defined on a box B with $\mathbf{f}(\mathbf{z}) \ne 0$ on the boundary of B, bound all the zeros of $\mathbf{f}$ in B by giving to k decimals

1. Points identified as simple zeros, or
2. Points identified as zeros of apparent multiplicity m greater than one.

A *zero of apparent multiplicity m greater than one* may be taken to mean a zero such that a topological degree computation over a small domain enclosing it yields the integer m.

     A procedure for solving this problem is easily obtained by generalizing the procedure described for solving the previous problem, where there was just one function $f(z)$. The topological degree computed for $\mathbf{f}(\mathbf{z})$, considered as $2n$ component elementary real functions of $2n$ real variables serves as before to isolate the various zeros. The topological degree cannot be negative as the well-known theorem below implies.

     THEOREM **6.3**: *For an elementary analytic function* $\mathbf{f}(\mathbf{z})$, *the real Jacobian* $\frac{\partial(u_1, v_1, \ldots, u_n, v_n)}{\partial(x_1, y_1, \ldots, x_n, y_n)}$ *equals the square of the absolute value of the complex Jacobian* $\frac{\partial(f_1, \ldots, f_n)}{\partial(z_1, \ldots, z_n)}$.

After the Cauchy-Riemann equations are used to replace all partial derivatives with respect to the variables y_k, the real Jacobian has the form

$$
\det
\begin{bmatrix}
\dfrac{\partial u_1}{\partial x_1} & \dfrac{\partial v_1}{\partial x_1} & \cdots & \cdots & \dfrac{\partial u_1}{\partial x_n} & \dfrac{\partial v_1}{\partial x_n} \\[2ex]
-\dfrac{\partial v_1}{\partial x_1} & \dfrac{\partial u_1}{\partial x_1} & \cdots & \cdots & -\dfrac{\partial v_1}{\partial x_n} & \dfrac{\partial u_1}{\partial x_n} \\[2ex]
\vdots & \vdots & \vdots\vdots & \vdots\vdots & \vdots & \vdots \\[2ex]
\dfrac{\partial u_n}{\partial x_1} & \dfrac{\partial v_n}{\partial x_1} & \cdots & \cdots & \dfrac{\partial u_n}{\partial x_n} & \dfrac{\partial v_n}{\partial x_n} \\[2ex]
-\dfrac{\partial v_n}{\partial x_1} & \dfrac{\partial u_n}{\partial x_1} & \cdots & \cdots & -\dfrac{\partial v_n}{\partial x_n} & \dfrac{\partial u_n}{\partial x_n}
\end{bmatrix}
\tag{6.25}
$$

Consider the matrix

$$
A = \frac{1}{\sqrt{2}}
\begin{bmatrix}
1 & 1 \\
i & -i
\end{bmatrix}
$$

with inverse

$$
A^{-1} = \frac{1}{\sqrt{2}}
\begin{bmatrix}
1 & -i \\
1 & i
\end{bmatrix}
$$

For any real constants a, b we have

$$
A^{-1}
\begin{bmatrix}
a & b \\
-b & a
\end{bmatrix}
A =
\begin{bmatrix}
a + ib & 0 \\
0 & a - ib
\end{bmatrix}
=
\begin{bmatrix}
a + ib & 0 \\
0 & a + ib
\end{bmatrix}
$$

If we premultiply the matrix of line 6.25 with a $2n$-square matrix composed of n submatrices A^{-1} in diagonal position and postmultiply with a $2n$-square matrix composed of n submatrices A in diagonal position, the determinant of the result equals the Jacobian 6.25 and has the form

$$\det \begin{bmatrix} \dfrac{\partial f_1}{\partial z_1} & 0 & \cdots & \cdots & \dfrac{\partial f_1}{\partial z_n} & 0 \\[2mm] 0 & \dfrac{\overline{\partial f_1}}{\partial z_1} & \cdots & \cdots & 0 & \dfrac{\overline{\partial f_1}}{\partial z_n} \\[2mm] \vdots & \vdots & \vdots\vdots\vdots & \vdots\vdots\vdots & \vdots & \vdots \\[2mm] \dfrac{\partial f_n}{\partial z_1} & 0 & \cdots & \cdots & \dfrac{\partial f_n}{\partial z_n} & 0 \\[2mm] 0 & \dfrac{\overline{\partial f_n}}{\partial z_1} & \cdots & \cdots & 0 & \dfrac{\overline{\partial f_n}}{\partial z_n} \end{bmatrix} \tag{6.26}$$

By performing a certain number of column exchanges on the matrix above followed by the same number of row exchanges, we can make the determinant take the form below.

$$\det \begin{bmatrix} \dfrac{\partial f_1}{\partial z_1} & \cdots & \dfrac{\partial f_1}{\partial z_n} & 0 & \cdots & 0 \\[2mm] \vdots & \vdots\vdots\vdots & \vdots & \vdots & \vdots\vdots\vdots & \vdots \\[2mm] \dfrac{\partial f_n}{\partial z_1} & \cdots & \dfrac{\partial f_n}{\partial z_n} & 0 & \cdots & 0 \\[2mm] 0 & \cdots & 0 & \dfrac{\overline{\partial f_1}}{\partial z_1} & \cdots & \dfrac{\overline{\partial f_1}}{\partial z_n} \\[2mm] \vdots & \vdots\vdots\vdots & \vdots & \vdots & \vdots\vdots\vdots & \vdots \\[2mm] 0 & \cdots & 0 & \dfrac{\overline{\partial f_n}}{\partial z_1} & \cdots & \dfrac{\overline{\partial f_n}}{\partial z_n} \end{bmatrix}$$

We see then that the real Jacobian equals the product of the complex Jacobian times its own conjugate or the square of the absolute value of the complex Jacobian, and so the real Jacobian can never be negative.

EXERCISES

1. Call `roots` and find to 10 decimals the zeros of the two polynomials $z^4 - 10z^3 + 35z^2 - 50z + 24$ and $z^4 - 10z^3 - 35z^2 - 50z - 24$ mentioned in Section 6.2.

2. Although the program `r_roots` determines the multiplicity of the zeros it finds rather than the *apparent multiplicity* like the programs `roots` or

c_roots, the Euclidean algorithm that is employed sometimes takes a lot of computer time. Call r_roots and find to 10 decimals the zeros of $0.456z^{10} + 32.56z^9 + 577.32z^8 + 56.97z^7 + 44.3z^6 + 22.53z^5 + 11.00001z^4 + 21.1115z^2 + 11.63z + 1231.01$, or a similar polynomial, and see how long the program takes. Then feed the problem to roots by calling roots < r_roots.log, and notice the improved solution time.

3. Call c_calc and find to 10 decimals the values of log 1, log i and log -1. Note the error message for the last entry. The program c_calc employs the interpretation of log z mentioned in Section 6.6, so log z is undefined on the negative real line.

4. Call c_calc and find to 10 decimals the values of $\sqrt{1}$, $\sqrt{i}$ and $\sqrt{-1}$. Note the error message for the last entry. The program c_calc evaluates $\sqrt{z}$ using the interpretation given in line 6.19.

5. Call c_calc and find to 10 decimals the values of $(-1)^{1/3}$, $0^{1/3}$ and $i^{1/3}$. Note the error message for the last entry. The program c_calc evaluates $z^{1/3}$ using the interpretation given in line 6.21.

6. Call c_fdzero and find to 10 decimals the zeros of

$$f_1(z_1, z_2) = z_1^2 - z_2$$
$$f_2(z_1, z_2) = z_2^2 - z_1$$

Use the search rectangle $(0 \pm 10) + i(0 \pm 10)$ for both z_1 and z_2. This problem is similar to the one given in Exercise 1 of Chapter 5 except that here the variables range over complex rectangles instead of over real intervals. Note the four solutions instead of the two obtained previously.

7. Call c_fdzero and find to 5 decimals the zeros of sin z^2 in the rectangle with opposite vertices $-4-i$ and $3 + 2i$. The last zero found is the zero at the origin with multiplicity 2. This zero is not located by Newton's method but by the slow method of repeated bisection of an enclosing rectangle. The delay in finding the last zero becomes more noticeable if 20 decimals are requested. Repeat the solution to 20 decimals by editing the file c_fdzero.log appropriately and using the command line c_fdzero < c_fdzero.log.

NOTES

N6.1. The use of the Euclidean algorithm to determine the multiplicity structure of a polynomial is discussed in Uspensky's textbook [6] and in the paper by Dunaway [3].

N6.2. Theorem 6.1 and 6.2 were discovered by Cauchy. Wilf's book [7] discusses these and other polynomial zero bounds.

N6.3. Our presentation of Bairstow's method follows that of Hamming [4].

N6.4. The error bound for polynomials used in Section 6.9 was given by Braess and Hadeler [1].

N6.5. The demonstration program `c_fdzero` computes the topological degree to determine the number of zeros in a search rectangle in the complex plane. The papers by Collins and Krandick [2] and by Schaefer [5] discuss using the argument principle for this purpose.

REFERENCES

1. Braess, D. and Hadeler, K. P., Simultaneous inclusion of the zeros of a polynomial, *Numerische Mathematik* **21** (1973), 161–165.

2. Collins, G. E. and Krandick, W., *An efficient algorithm for infallible polynomial complex root isolation*, Proceedings of ISSAC, 1992.

3. Dunaway, D. K., Calculation of zeros of a real polynomial through factorization using Euclid's algorithm, *SIAM J. Numer. Anal.* **11** (1974), 1087–1104.

4. Hamming, R. W., *Numerical Methods for Scientists and Engineers, 2nd Edit.*, McGraw-Hill, New York, 1973.

5. Schaefer, M. J., Precise zero of analytic functions using interval arithmetic, *Interval Computations* **4** (1993), 22–39.

6. Uspensky, J. V., *Theory of equations*, McGraw-Hill, New York, 1948.

7. Wilf, H. S., *Mathematics for the physical sciences*, John Wiley & Sons, New York, 1962.

VII

PROBLEMS OF LINEAR ALGEBRA

A linear algebra computation problem in general requires that one or more matrices be specified to define the problem's initial conditions. A matrix is called *real, complex, rational,* or *complex rational* if all its elements are, respectively, real, complex, rational, or complex rational. We consider first problems in which all initial matrices are real but not necessarily rational, or complex but not necessarily complex rational. Perhaps a matrix element is specified in terms of standard functions, for example, as $\tan^{-1}\frac{\pi}{5}$, or perhaps it requires a separate subprogram for its computation. We can compute any matrix element as accurately as we please, but we will have the usual difficulty in deciding whether the element is or is not equal to another number. Later in the chapter we consider linear algebra problems in which all initial matrices are rational or complex rational. Here, besides range arithmetic, we can use rational arithmetic in our solution procedure, and this improves the results we can obtain.

When a theorem appears in this chapter without any proof, it is a linear algebra result that may not be covered in an elementary linear algebra course. The linear algebra references given in the Notes section may be consulted for a proof.

7.1. NOTATION

Generally for a matrix A of m rows and n columns, or, more briefly, of size $m \times n$, we indicate its components a_{ij} by a small letter with row and column indices. We try to match the small letter with the capital letter.

An *upper triangular* matrix is an n-square matrix of the form

$$\begin{bmatrix} a_{11} & a_{12} & a_{13} & \cdots & a_{1n} \\ & a_{22} & a_{23} & \cdots & a_{2n} \\ & & a_{33} & \cdots & a_{3n} \\ & & & \ddots & \vdots \\ & & & & a_{nn} \end{bmatrix}$$

An entirely blank region of a displayed matrix, as occurs above, is always to be understood as made up entirely of zero elements. Thus an upper triangular matrix can be characterized as a square matrix whose elements below the diagonal are all zero. The diagonal elements of any matrix, square or not, are those elements with matching row and column indices. The determinant of an upper triangular matrix equals the product of its diagonal elements. Similarly, a lower triangular matrix is an n-square matrix with zero elements above its diagonal, and its determinant also is equal to the product of its diagonal elements.

For any matrix A, the symbol $|A|$ denotes a matrix of the same size as A with its elements equal to $|a_{ij}|$. A matrix inequality $A \leq B$ implies that the two matrices A and B are the same size and that $a_{ij} \leq b_{ij}$ for all i, j. The matrix inequality $A < B$ has an analogous interpretation.

We use the symbol I to denote an n-square identity matrix, its size inferred from context, and the symbol O to denote an $m \times n$ matrix with all elements zero, its size also inferred from context. If there is only one column to a matrix, it is customary to omit the unchanging column index for its elements and to call the matrix a *vector*. Its elements may then also be called *components*. We use lower case boldface letters to denote vectors. So if $\mathbf{c}$ is a vector, it has the form

$$\begin{bmatrix} c_1 \\ c_2 \\ \vdots \\ c_n \end{bmatrix}$$

The vector $\mathbf{c}$ may also be called an n-vector to indicate the number of its elements. The length of $\mathbf{c}$, denoted by len $\mathbf{c}$, is $\sqrt{\sum_{i=1}^{n} c_n^2}$ if $\mathbf{c}$ is real and is $\sqrt{\sum_{i=1}^{n} |c_n|^2}$ if $\mathbf{c}$ is complex.

The transpose of a matrix A of size $m \times n$, denoted by the symbol A^T, is a matrix D of size $n \times m$ with d_{ij} equal to a_{ji}. The rule $(AB)^T = B^T A^T$ is true if the product AB is defined. The transpose $\mathbf{c}^T$ of an n-vector $\mathbf{c}$ is then a one row matrix with n elements.

7.2. SOLVING n LINEAR EQUATIONS IN n UNKNOWNS

A frequently encountered problem is that of solving the system of linear equations below for the values of $x_1, x_2, \ldots, x_n$.

$$
\begin{aligned}
a_{11}x_1 + a_{12}x_2 + \cdots + a_{1n}x_n &= b_1 \\
a_{21}x_1 + a_{22}x_2 + \cdots + a_{2n}x_n &= b_2 \\
\vdots \qquad \vdots \qquad\qquad \vdots \qquad &\ \ \vdots \\
a_{n1}x_1 + a_{n2}x_2 + \cdots + a_{nn}x_n &= b_n
\end{aligned}
$$

This is equivalent to solving the matrix equation

$$A\mathbf{x} = \mathbf{b}$$

for $\mathbf{x}$, where A is the n-square coefficient matrix

$$
\begin{bmatrix}
a_{11} & a_{12} & \cdots & a_{1n} \\
a_{21} & a_{22} & \cdots & a_{2n} \\
\vdots & \vdots & \vdots & \vdots \\
a_{n1} & a_{n2} & \cdots & a_{nn}
\end{bmatrix}
$$

and $\mathbf{x}$ and $\mathbf{b}$ are vectors with elements $x_1, \ldots, x_n$, and $b_1, \ldots, b_n$, respectively.

Often the goal here is

NONSOLVABLE PROBLEM 7.1: Given a set of n equations in n unknowns, represented by the matrix equation $A\mathbf{x} = \mathbf{b}$, where A is a real (complex) n-square coefficient matrix, $\mathbf{x}$ is the vector of unknowns, and $\mathbf{b}$ the real (complex) vector of constants, find to k decimals the unknowns $x_1, \ldots, x_n$, or else indicate that $\det A = 0$.

Of course the difficulty here is the necessity of determining whether or not $\det A = 0$, an instance of Nonsolvable Problem 3.1. To get a solvable problem, we give up trying to determine whether the A determinant is *exactly* zero.

SOLVABLE PROBLEM 7.2: Given a set of n equations in n unknowns, represented by the matrix equation $A\mathbf{x} = \mathbf{b}$, where A is a real (complex) n-square coefficient matrix, $\mathbf{x}$ is the vector of unknowns, and $\mathbf{b}$ the real (complex) vector of constants, find to k decimals the unknowns $x_1, \ldots, x_n$, or else indicate that $|\det A| < 10^{-k}$.

The bound 10^{-k} is merely a convenient one and could be replaced by another value, for instance, 10^{-nk} or 10^{-k^2}.

The equations can be solved by the elimination method, whereby the equations are manipulated until they have the matrix representation $A'\mathbf{x} = \mathbf{b}'$, where A' is upper triangular. If we are successful in obtaining this representation for the equations, then they have the form

$$a'_{11}x_1 + a'_{12}x_2 + \cdots + a'_{1n}x_n = b'_1$$
$$a'_{22}x_2 + \cdots + a'_{2n}x_n = b'_2$$
$$\vdots \qquad \qquad \vdots$$
$$a'_{nn}x_n = b'_n$$

and are easily solved by back substitution, namely

$$x_n = \frac{1}{a'_{nn}}(b'_n)$$

$$x_{n-1} = \frac{1}{a'_{n-1,n-1}}(b'_{n-1} - a'_{n-1,n}x_n)$$

$$\vdots \qquad \vdots$$

$$x_1 = \frac{1}{a'_{11}}(b'_1 - a'_{12}x_2 - \cdots - a'_{1n}x_n)$$

The equations above can be combined as

$$x_i = \frac{1}{a'_{ii}}\left(b'_i - \sum_{k=i+1}^{n} a'_{ik}x_k\right) \qquad i = n, n-1, \ldots, 2, 1 \qquad (7.1)$$

with the understanding that the sum is void for x_n because here the index k runs from $n + 1$ up to n.

The manipulation of the equations is done most conveniently by repeatedly changing the stored representations of A and $\mathbf{b}$ until the desired upper triangular form of A is obtained. The operations performed on these matrices are of just two types:

Exchange row(i,j) Exchange rows i and j

Add row multiple(i,j,m) And m times row i to a different row j

Either matrix operation always is performed simultaneously on A and on $\mathbf{b}$. When either operation is executed, the set of linear equations represented by $A\mathbf{x} = \mathbf{b}$ is changed by an allowable algebraic manipulation into a different set of equations. The two sets of equations are equivalent, because it is possible to undo each row operation by another operation of the same type.

The upper triangular form of A is obtained by bringing the columns of A to the proper form, one by one, starting at column 1 and working toward the right. For the sake of simplicity, we use a'_{ij} to denote an element of A during this process, and we do not try to differentiate between the various successive values that the element may take. When working on column j, the procedure is as follows: The elements $a'_{jj}, a'_{j+1,j}, \ldots, a'_{nj}$ that test $\neq 0$ are examined to locate one with absolute value $\geq$ the others. If this is a'_{kj} and $k \neq j$, then the operation exchange row(j, k) is performed to make this element a'_{jj}. Then using the other row operation, appropriate multiples of row j are subtracted from rows below to clear the A elements in column j below a'_{jj}. Doing the clearing with the A element in column j having the largest absolute value, rather than using any element that tests $\neq 0$ in row j or below, has the advantage that smaller multiples of row j are subtracted from rows below, and this makes the A elements to the right of column j tend to have smaller wids.

If we are successful in bringing A to upper triangular form, then we obtain the values of the variables x_i by using Equation 7.1. We determine whether k correct decimals have been obtained, and if not, we increase precision an appropriate amount and the entire computation is repeated.

Let us be more specific about the phrase "increase precision an appropriate amount," which appears in the preceding paragraph and at other places in this chapter. The number of correct decimals, in either fixed-decimal or scientific floating-point notation, that can be obtained from a ranged number is determinable from the number's mid and wid. By examining all the computed answers of a problem, it is possible to determine whether the target number of correct decimals has been achieved, and, if not, what the maximum deficit is correct places is. In Chapter 10, the particular functions for making these determinations are described. Normally when a computation procedure is redone because answers are not accurate enough, precision is increased by the maximum deficit.

It may happen when working with a certain column j of A that all the elements $a'_{jj}, a'_{j+1,j}, \ldots, a'_{nj}$, test $\doteq 0$. In this case we abandon the attempt to solve the equations, and compute det A instead. Any operation add row multiple (i, j, m) that has been performed has not changed the determinant of our varying matrix A. Any operation exchange row (i, j) has changed just the sign of this determinant. Now a third operation is allowed:

$$\text{Exchange col}(i,j) \qquad \text{exchange columns } i \text{ and } j$$

This operation also changes just the sign of det A. Our changed procedure now is to ignore **b** but to continue trying to bring A to upper triangular form, using, if necessary, this additional operation. Now when dealing with any column j of A, if in row j or below we can find no elements that test $\neq 0$, we search the columns of A further to the right for an element $\neq 0$ in row j or below. If one is found in column k, then we perform the operation exchange col(j, k) to bring this element into column j, and then continue the process just described of making this element a'_{jj} and using it to clear all elements below it. The procedure now terminates when for some column q, no element that tests $\neq 0$ in row q or below is found in this column or in the other A columns further to the right. The final A matrix has the form

$$A' = \begin{bmatrix} a'_{11} & a'_{12} & \cdots & a'_{1q} & \cdots & a'_{1n} \\ & a'_{22} & \cdots & a'_{2q} & \cdots & a'_{2n} \\ & & \ddots & \vdots & \vdots\vdots\vdots & \vdots \\ & & & a'_{qq} & \cdots & a'_{qn} \\ & & & \vdots & \vdots\vdots\vdots & \vdots \\ & & & a'_{nq} & \cdots & a'_{nn} \end{bmatrix}$$

where the submatrix

$$\begin{bmatrix} a'_{qq} & \cdots & a'_{qn} \\ \vdots & \vdots\vdots\vdots & \vdots \\ a'_{nq} & \cdots & a'_{nn} \end{bmatrix}$$

has all elements testing $\doteq 0$. The determinant of the matrix A' equals the product of the diagonal elements $a'_{11}, \ldots, a'_{q-1,q-1}$ times the determinant of the submatrix and is identical to the determinant of the starting matrix A, except that its sign may be wrong. In magnitude, the determinant of the submatrix is bounded above by the product of magnitude bounds for its various columns, times the number of terms in the determinant, which is $(n - q + 1)!$. The element a'_{ij} has the magnitude bound $|\text{mid } a'_{ij}| + \text{wid } a'_{ij}$, so by taking the maximum of these for each column, multiplying these maxima together, and then multiplying by $(n - q + 1)!$, we obtain a magnitude bound M for the determinant of the submatrix. We then obtain for $|\det A|$ the ranged value

$$|\det A| = 0 \oplus |a'_{11} \cdots a'_{q-1,q-1}| M$$

Now we can determine whether we can use the escape of Solvable Problem 7.2, namely $|\det A| \leq 10^{-k}$. If we cannot, then we try another cycle of the entire evaluation process at an appropriate higher precision.

7.3. SOLVING n LINEAR INTERVAL EQUATIONS IN n UNKNOWNS

Suppose that n linear equations in n unknowns are given

$$\begin{aligned} a_{11}x_1 + a_{12}x_2 + \cdots + a_{1n}x_n &= b_1 \\ a_{21}x_1 + a_{22}x_2 + \cdots + a_{2n}x_n &= b_2 \\ \vdots \quad\quad \vdots \quad\quad\quad\quad \vdots \quad\quad \vdots \\ a_{n1}x_1 + a_{n2}x_2 + \cdots + a_{nn}x_n &= b_n \end{aligned} \tag{7.2}$$

and that the elements a_{ij} and b_i are not known exactly but are specified as intervals. This situation arises sometimes when these numbers are experimentally determined, being known only to a few correct decimal places. For instance, we may have the equations

$$2.23\tilde{\ }x_1 + 1.73\tilde{\ }x_2 - 2.70\tilde{\ }x_3 = 4.19\tilde{\ }$$

$$-1.62\tilde{\ }x_1 + 6.77\tilde{\ }x_2 + 3.45\tilde{\ }x_3 = 1.28\tilde{\ } \qquad (7.3)$$

$$4.16\tilde{\ }x_1 + 2.81\tilde{\ }x_2 + 3.93\tilde{\ }x_3 = 6.16\tilde{\ }$$

where all constants are known only to 2 correct decimals. Thus a_{11} lies in the interval 2.23 ± 0.005, or using the endpoint format, the interval [2.225, 2.235]. A solution to these equations is required with all the constants replaced by appropriate intervals, the determined value for each variable x_i being an interval also.

This problem can also occur with specific intervals prescribed in advance for all constants a_{ij} and b_i. Thus we may have the equations

$$[1, 1.5]x_1 + [e^{1.1}, e^{1.2}]x_2 = [3, 4]$$

$$[2, 3]x_1 + [-4.5, -4.4]x_2 = [9, 9]$$

Here the constant b_2 equals 9 exactly.

If $\hat{A}$ is a real $n \times n$ matrix with each element $\hat{a}_{ij}$ somewhere within the corresponding interval a_{ij}, and, similarly, $\hat{\mathbf{b}}$ is a real n-vector with each component $\hat{b}_i$ within b_i, then the matrix equation $\hat{A}\mathbf{x} = \hat{\mathbf{b}}$ has a unique solution vector $\mathbf{x}$ if the matrix $\hat{A}$ is nonsingular. The vector $\mathbf{x}$ defines a solution point in n-dimensional Euclidean space, and similarly the collection of vectors $\mathbf{x}$ obtained as $\hat{A}$ and $\hat{\mathbf{b}}$ range over their possibilities defines a connected set of solution points in this space. One method of 'solving' the system of interval equations 7.2 is to provide an optimum finite interval bound for each component of $\mathbf{x}$ whenever such bounds exist, and otherwise, when some matrix $\hat{A}$ is singular, to then identify the interval matrix A as singular. The interval bounds for the components x_i determine an n-dimensional box, with sides perpendicular to the coordinate axes, within which the set of solutions must lie. In general, not all points inside this box are solution points, but the size of the box cannot be reduced.

In Section 7.15 where we study this problem in more detail, we obtain

SOLVABLE PROBLEM 7.18: Given a system of n linear interval equations in n unknowns, represented by the matrix equation $Ax = \mathbf{b}$ with A and $\mathbf{b}$ having interval elements with real endpoints, give to k decimals optimum upper and lower bounds for the unknowns x_i, indicate that the interval matrix A is singular, or indicate that $|\det A| < 10^{-k}$ when all A elements are set to the right endpoints of their intervals.

A simple way of obtaining interval bounds for x_i is to follow the elimination procedure of the preceding section, omitting the determinant computation part, but with formal interval arithmetic in place of the range arithmetic. The procedure comes to a successful conclusion if for each column j we are able to find, in row j or below, an interval element a'_{ij} not containing 0. In column j, if there are several such interval elements, it is generally best to choose for the clearing operation the one with the smallest ratio wid $a'_{kj}/|\text{mid } a'_{kj}|$, because this choice generally leads to narrower intervals to the right of column j.

Although this interval elimination procedure is often successful, the interval values obtained for x_i are usually considerably wider than the optimum intervals. In Table 7.1, the Method 1 column gives the results of applying the interval elimination procedure to the interval equation system 7.3. The correct (to four decimals) intervals for x_i are given in the rightmost column and are narrower. For larger systems of interval equations, the overestimation of the x_i intervals using this first method is often more serious.

When performing any of the four operations, $+$, $-$, $\times$, $\div$, on intervals, an overly large result interval may be obtained if the operands are related in some way. Each of the four interval arithmetic relations (4.24)–(4.27) given in Chapter 4 is obtained under the assumption that the two operands are independent, so that if one operand takes a certain value within its interval, this has no influence on the other operand. If this assumption is not valid, a narrower result interval may be possible.

Two simple examples of this are the computations of $x - x$ and $xy - x$ for any nonpoint intervals x and y. For instance, if $x = 2 \pm 1$ and $y = 3 \pm 1$, then according to Equation 4.25, we have $x - x = 0 \pm 2$, when the answer should be 0 ± 0. Similarly, according to Equations 4.25 and 4.26, we have

$$xy - x = (2 \pm 1) \times (3 \pm 1) - (2 \pm 1)$$

$$= (2 \cdot 3 \pm [1 \cdot 3 + 2 \cdot 1 + 1 \cdot 1]) - (2 \pm 1)$$

$$= (6 \pm 6) - (2 \pm 1) = 4 \pm 7$$

Here as x varies in the interval $[1, 3]$ and y varies in the interval $[2, 4]$, the function $xy - x$ varies in $[1, 9]$, and so the best possible mid-wid result is 4 ± 5.

TABLE 7.1 Interval Solutions to the Interval Equation System 7.3 Obtained Three Ways

	Method 1	Method 2	Optimum
x_1	[1.2325, 1.2607]	[1.2328, 1.2604]	[1.2415~, 1.2517~]
x_2	[0.5609, 0.5752]	[0.5610, 0.5750]	[0.5640~, 0.5721~]
x_3	[−0.1643, −0.1523]	[−0.1641, −0.1525]	[−0.1630~, −0.1536~]

When performing the elimination procedure on the interval matrix A and the interval vector $\mathbf{b}$, accurate results are obtained when doing the clearing of column 1 of A, because no relations exist among the starting elements a_{ij}, but when we clear other columns, the elements a'_{ij} found in the column have become linear combinations of the original column elements, and now are related. So overestimation of result intervals begins in column 2 and becomes progressively more serious as other columns are cleared. We describe next a method of improving the accuracy of interval arithmetic, which has been suggested by Hansen [8, 9].

7.4. DEGREE I INTERVAL ARITHMETIC

Suppose a certain interval computation depends on N initial constants with interval values. We imagine the i-th such interval, $m_i \pm w_i$, defining a variable v_i in $[m_i - w_i, m_i, + w_i]$. When we do an interval computation C involving these variables v_i, we obtain an interval result that may be considered a degree 0 series expansion of C in terms of the variables v_i. Suppose we express C as a degree 1 series expansion of the variables v_i and obtain interval bounds for all of C's degree 1 terms. This type of computation occurred in Chapter 4 where we needed to bound a high-degree term of a function $f(x)$ being expressed as a power series. To obtain these interval bounds, we use formal interval arithmetic, and each primitive variable v_i becomes

$$v_i = (m_i \pm w_i) + (1 \pm 0)(v_i - m_i) \tag{7.4}$$

The computation C proceeds by generating a degree 1 series representation for each successive quantity occurring in the calculation. In general an intermediate result q has the form

$$q = \alpha + \sum_k \beta_k(v_k - m_k) \tag{7.5}$$

Here the interval α defines q's degree 0 term, and an interval coefficient β_k is present to define a degree 1 term for each variable v_k on which q depends. The degree 0 term gives q's value as computed by ordinary interval arithmetic. The degree 1 terms give interval bounds for q's partial derivatives with respect to the various variables v_i. The quantity q with representation 7.5 may be viewed as a function $f(v_1, \ldots, v_N)$ of the N variables $v_1, \ldots, v_N$. The domain of this function is the N-space box B defined by the N intervals $m_i \pm w_i$, and the function f has the Taylor series expansion

$$f(m_1, \ldots, m_N) + \sum_k \frac{\partial f(m_1, \ldots, m_N)}{\partial v_k}(v_k - m_k) + \cdots$$

with the series expansion point $(m_1, \ldots, m_N)$ in B, so that $f(m_1, \ldots, m_N)$

equals the midpoint of q's interval α, and $\frac{\partial f(m_1, \ldots, m_N)}{\partial v_k}$ equals the midpoint of the interval β_k. For any point $v_1, \ldots, v_N$ within B, we have the interval relation (see Equation 4.40)

$$f(v_1, \ldots, v_N) = \text{mid } \alpha + \sum_k \beta_k(v_k - m_k)$$

The maximum absolute value of $(v_k - m_k)$ is w_k, so we obtain a second halfwidth for q's interval, namely

$$\text{wid } q = \sum_k |\beta_k| w_k \tag{7.6}$$

Here $|\beta_k|$ equals $|\text{mid } \beta_k| + \text{wid } \beta_k$. Thus we always have two wids for q, one by ordinary interval arithmetic (which we call the degree 0 wid) and one by using the relation 7.6 (which we call the degree 1 wid). In a computation, the degree 1 wid is not automatically computed as is the degree 0 wid, but whenever it is computed, it can replace q's degree 0 wid if it is smaller.

Repeating the computations given earlier for $x - x$ and $xy - x$ in this new format, if x is v_1 and y is v_2, we have

$$x - x = [(2 \pm 1) + (1 \pm 0) \cdot (v_1 - 2)] - [(2 \pm 1) + (1 \pm 0) \cdot (v_1 - 2)]$$

$$= [(0 \pm 2) + (0 \pm 0) \cdot (v_1 - 2)]$$

and

$$xy - x = [(2 \pm 1) + (1 \pm 0) \cdot (v_1 - 2)] \cdot [(3 \pm 1) + (1 \pm 0) \cdot (v_2 - 3)] -$$

$$-[(2 \pm 1) + (1 \pm 0) \cdot (v_1 - 2)]$$

$$= [(6 \pm 6) + (3 \pm 1) \cdot (v_1 - 2) + (2 \pm 1) \cdot (v_2 - 3)] -$$

$$-[(2 \pm 1) + (1 \pm 0) \cdot (v_1 - 2)]$$

$$= [(4 \pm 7) + (2 \pm 1) \cdot (v_1 - 2) + (2 \pm 1) \cdot (v_2 - 3)]$$

For $x - x$ we obtain a degree 1 wid of $0 \cdot 1 = 0$ and for $xy - x$ a degree 1 wid of $3 \cdot 1 + 3 \cdot 1 = 6$, an improvement over the degree 0 wids of 2 and 7, respectively. We see that now we obtain the correct wid for the first example and an improved but not minimum wid for the second example.

In general, degree 1 interval arithmetic requires more computation than the ordinary variety, but the wids it obtains often are significantly better for cases where the computation requires many additions and subtractions of intermediate results. Partial derivative bounds are obtained as a by-product, and these are useful sometimes, occurring in the application described in the next section.

The degree 1 wid can also be *larger* than the degree 0 wid. For instance, using the x and y intervals that have appeared in our examples, for $x \cdot y$ we obtain

$$x \cdot y = [(2 \pm 1) + (1 \pm 0) \cdot (v_1 - 2)] \cdot [(3 \pm 1) + (1 \pm 0) \cdot (v_2 - 3)]$$

$$= [(6 \pm 6) + (3 \pm 1) \cdot (v_1 - 2) + (2 \pm 1) \cdot (v_2 - 3)]$$

For $x \cdot y$ we obtain a degree 1 wid of $4 \cdot 1 + 3 \cdot 1 = 7$, which is larger than the degree 0 wid of 6. The situation is similar for the division x/y.

One can contemplate obtaining a ''degree 2 wid'' also. Here we have the same representation for all starting constants as before, but any quantity q computed in terms of these is calculated up to degree 2. Thus q has the form

$$q = \alpha + \sum_k \beta_k(v_i - m_k) + \sum_{k,l} \gamma_{kl}(v_k - m_k)(v_l - m_l)$$

Using similar reasoning as was used with the degree 1 case, the degree 2 wid obtained for q is

$$\text{wid } q = \sum_k |\text{mid } \beta_k| w_k + \sum_{k,l} |\gamma_{kl}| w_k w_l$$

For the example $xy - x$ we obtain now

$$xy - x = [(2 \pm 1) + (1 \pm 0) \cdot (v_1 - 2)] \cdot [(3 \pm 1) + (1 \pm 0) \cdot (v_2 - 3)] -$$

$$-[(2 \pm 1) + (1 \pm 0) \cdot (v_1 - 2)]$$

$$= [(4 \pm 7) + (2 \pm 1) \cdot (v_1 - 2) + (2 \pm 1) \cdot (v_2 - 3)$$

$$+ (1 \pm 0)(v_1 - 2)(v_2 - 3)]$$

The degree 2 wid is $[(2 \cdot 1) + (2 \cdot 1)] + [1 \cdot 1 \cdot 1] = 5$, which is the optimum value.

If there are N initial constants, the number of terms an end result must carry to obtain a degree 2 wid is proportional to N^2, while the number of terms for a degree 1 wid is proportional only to N, and this increased computational cost makes it difficult to find applications where it is worthwhile computing degree 2 wids instead of degree 1 wids.

7.5. SOLVING INTERVAL LINEAR EQUATIONS WITH DEGREE 1 INTERVAL ARITHMETIC

Optimum interval bounds for the unknowns can be obtained using degree 1 arithmetic. The method does not work in all cases, as does an alternate method using rational arithmetic described in Section 7.15, but the method is applicable to commonly encountered cases such as the equation set 7.3, where the matrix A and vector $\mathbf{b}$ intervals are not overly wide.

All the elements a_{ij} and b_i on which the solution depends, previously regarded as initial constants, now become variables of the form 7.4. We have then $n^2 + n$ different variables v_i. The elimination procedure of Section 7.2 is used to obtain degree 1 expressions for all the variables x_i. As mentioned previously, with this type of interval arithmetic, the degree 1 wid 7.6 is not obtained automatically as is the degree 0 wid. So it is a matter of programming when to obtain this wid, which replaces the degree 0 wid if it is smaller. Because the A matrix elements used for clearing a column play a crucial role, a degree 1 wid is computed for them and for the final x_i values. The intervals of the middle column of Table 7.1 were obtained following this plan and are somewhat smaller than the degree 0 wids of the first column.

To obtain the optimum interval for a particular variable x_s, we examine its degree 1 terms. We can expect, because x_s depends on all elements a_{ij} and b_i, that there is a term in $(v_k - m_k)$ in x_s's degree 1 representation corresponding to each element of A and to each component of $\mathbf{b}$. If β_k is the interval coefficient of the term corresponding to a_{ij}, then $\partial x_s/\partial a_{ij}$ is always some point in β_k for any choice of elements of A or components of $\mathbf{b}$ within their respective intervals. Or if β_k is the interval coefficient of the term corresponding to b_i, then $\partial x_s/\partial b_i$ is always a point in β_k.

If all these $n^2 + n$ intervals β_k appearing in the expression for x_s do not contain zero, then we have a means of obtaining the optimum interval bounds for x_s. To obtain the largest possible value for x_s, we set each element a_{ij} (component b_i) to its right interval endpoint or to its left interval endpoint depending on whether the β_k interval corresponding to the element (component) is positive or negative. We then solve the resulting (noninterval) linear equations, obtaining this way the largest possible value for x_s. To obtain the smallest possible value for x_s, we set each element a_{ij} and component b_i to its other endpoint and again solve the resulting linear equations. Thus to obtain the largest and smallest possible values for all the variables x_s, we need to solve $2n$ different sets of linear equations. The third column of entries in Table 7.1 is obtained this way.

This method fails if any of the intervals β_k contain 0, and in such cases we need to use the more complicated procedure described in Section 7.15.

7.6. FINDING THE INVERSE OF A REAL OR COMPLEX SQUARE MATRIX

Suppose for a square matrix A we require the inverse matrix A^{-1} satisfying the equation $AA^{-1} = A^{-1}A = I$. The inverse matrix A^{-1} exists if $\det A \neq 0$. The appropriate objective is

SOLVABLE PROBLEM 7.3: Given the n-square real (complex) matrix A, find to k decimals the elements of the matrix A^{-1}, or else indicate that $|\det A| \le 10^{-k}$.

We start with the matrix equation $AA^{-1} = I$, where the matrices A and I are known, and A^{-1} is unknown, and apply simultaneously to the matrices A and I a series of the operations exchange row (i, j) and add row multiple (i, j, m), attempting with these row operations to bring A into upper triangular form. An operation exchange row (i, j) on an n-square matrix C is equivalent to premultiplying C by a matrix equal to the identity matrix I except that rows i and j have been exchanged. Similarly an operation add row multiple (i, j, m) on C is equivalent to premultiplying C by a matrix equal to the identity matrix I except that m times row i has been added to row j. Thus we can justify performing either operation simultaneously on A and on I as simply the premultiplication of the equation $AA^{-1} = I$ by another matrix to produce the equation $A'A^{-1} = I'$.

Following the same procedure of gradually changing A that was described in Section 7.2, except that here we operate on A and I instead of A and **b**, we may or may not succeed in bringing A to upper triangular form. If we are unable to bring A to upper triangular form, then we abandon the goal of determining A^{-1} and instead obtain a bound on $|\det A|$ in the same way as described in section 7.2. If we succeed in bringing A to upper triangular form, then we begin to use a new row operation:

Multiply row(i,m) Multiply row i by the nonzero constant m.

Working from column n of A backward toward column 1, in column j we apply the operation multiply row $(j, 1/a'_{jj})$ to both A' and I' to make a'_{jj} equal 1. Again we can justify applying an operation multiply row (i, m) on A' and I' as equivalent to the premultiplication of the equation $A'A^{-1} = I'$ by a matrix equal to the identity matrix except that row i has been multiplied by the constant m. The next part of the column j procedure is to apply a series of operations add row multiple (i, j, m) to both A' and I' to clear the elements of A' above the unit element.

After all columns of A' have been processed, the matrix A' has been converted to the identity matrix I, and the matrix equation that initially was $AA^{-1} = I$ has been converted to $IA^{-1} = I'$. The elements of I' now equal the elements of A^{-1}. If this process leads to results that are insufficiently accurate, the entire calculation is repeated at an appropriate higher precision.

7.7. DETERMINING THE RANK OF A MATRIX

Let A be a real or complex matrix of size $m \times n$. The columns of A can be used to define n vectors, called *column vectors* of A, each vector having m components.

The *column rank* of A equals the maximum number of linearly independent column vectors of A, and accordingly the column rank cannot exceed n. Similarly, the rows of A can be used to define m vectors, called *row vectors* of A, each vector having n components. The *row rank* of A equals the maximum number of linearly independent row vectors of A, and the row rank cannot exceed m.

THEOREM 7.1: *The row rank of a real (complex) matrix A equals its column rank.*

As a consequence of this theorem, it is acceptable to use the term *rank* of a matrix to indicate either its row rank or its column rank. It is clear that the rank of A cannot exceed min (m, n).

Consider the three matrix operations used earlier for solving linear equations: exchange row (i, j), add row multiple (i, j, m), and exchange col (i, j). Applying either of the row operations to a matrix A clearly does not change its row rank, and applying the column operation also does not change the row rank because the operation only permutes the row vector components. To determine the row rank of a matrix A, we can apply these three operations in the way described in Section 7.2 and attempt to bring A to a kind of upper triangular form, that is, attempt to make zero all elements below the diagonal, getting as many rows of A as possible into the desired form. If $m < n$, the final form we achieve can be

$$A' = \begin{bmatrix} a'_{11} & a'_{12} & \cdots & a'_{1m} & \cdots & a'_{1n} \\ & u'_{22} & \cdots & a'_{2m} & \cdots & a'_{2n} \\ & & \ddots & \vdots & \vdots\vdots\vdots & \vdots \\ & & & a'_{mm} & \cdots & a'_{mn} \end{bmatrix}$$

with all diagonal elements $\neq 0$. Here the rank of A is m because the row vectors of A' are linearly independent. Similarly, if $m \geq n$, the final form can be

$$A' = \begin{bmatrix} a'_{11} & a'_{12} & \cdots & a'_{1n} \\ & a'_{22} & \cdots & a'_{2n} \\ & & \ddots & \vdots \\ & & & a'_{nn} \\ & & & 0 \\ & & & \vdots \\ & & & 0 \end{bmatrix}$$

with again diagonal elements $\neq 0$. Here the rank of A is n. If during the clearing process we encounter a column where no element $\doteq 0$ is found that can take the diagonal position, then the final form we achieve is

$$A' = \begin{bmatrix} a'_{11} & a'_{12} & \cdots & a'_{1r} & a'_{1,r+1} & \cdots & a'_{1n} \\ & a'_{22} & \cdots & a'_{2r} & a'_{2,r+1} & \cdots & a'_{2n} \\ & & \ddots & \vdots & \vdots & \vdots\vdots\vdots & \vdots \\ & & & a'_{rr} & a'_{r,r+1} & \cdots & a'_{rn} \\ & & & & a'_{r+1,r+1} & \cdots & a'_{r+1,n} \\ & & & & \vdots & \vdots\vdots\vdots & \vdots \\ & & & & a'_{m,r+1} & \cdots & a'_{mn} \end{bmatrix} \tag{7.7}$$

where the submatrix

$$\begin{bmatrix} a'_{r+1,r+1} & \cdots & a'_{r+1,n} \\ \vdots & \vdots\vdots\vdots & \vdots \\ a'_{m,r+1} & \cdots & a'_{mn} \end{bmatrix}$$

has all elements $\doteq 0$. This time the rank of A is in doubt. The rank is r if all the elements in the submatrix are actually zero, but the rank can be any integer between r and $\min(m, n)$ according as various elements in the submatrix turn out not to be zero. A recomputation of A' at a higher precision may or may not resolve the dilemma. It is only when the rank is found to equal $\min(m, n)$ that the rank is unambiguous.

The difficulty cannot be avoided by using some other, possibly better, method of determining rank, because in the case of a matrix A where it is known in advance that all off-diagonal elements are zero, determining rank becomes purely a matter of determining which diagonal elements are zero, a nonsolvable problem. Accordingly, we have

NONSOLVABLE PROBLEM 7.4: Determine the rank of a real (complex) matrix.

7.8. FINDING A SOLUTION TO Ax = 0 WHEN det A = 0

Often a system of n linear homogeneous equations in n unknowns is encountered

$$\begin{aligned} a_{11}x_1 + a_{12}x_2 + \cdots + a_{1n}x_n &= 0 \\ a_{21}x_1 + a_{22}x_2 + \cdots + a_{2n}x_n &= 0 \\ \vdots \qquad \vdots \qquad\qquad \vdots \qquad \vdots \\ a_{n1}x_1 + a_{n2}x_2 + \cdots + a_{nn}x_n &= 0 \end{aligned} \tag{7.8}$$

with the determinant of the coefficient matrix A known to equal zero, and a nontrivial solution is required. That is, an n-vector x, not equal to 0, is to be found satisfying the matrix equation $Ax = 0$. In the next section we see that this problem arises in finding eigenvectors.

If **x** is a solution vector, then so is c**x** for any choice of the constant c. Thus if we can find a nonzero solution vector, then we can also find one of length 1, namely $\frac{1}{\text{len } \mathbf{x}}$ **x**. Accordingly we take the problem now to be that of locating a solution vector of length 1. (An equally convenient choice would be a vector with its largest component equal to 1. A "vector with largest component 1" may be substituted for a "vector of length 1" in the various solvable and nonsolvable problems of this chapter.)

The procedure described earlier for solving linear equations can be adapted to solve this problem too. We start with the matrix equation $A\mathbf{x} = \mathbf{0}$, and apply the operations exchange row (i, j) and add row multiple (i, j, m) to bring A to upper triangular form. This time we can dispense with a column of constants, the **b** vector, because it is initially **0**, and these two operations never change it.

Because det $A = 0$, the rank of A is less than n, and eventually we encounter a column j of A where we find no elements in row j or below which test $\neq 0$. When this happens, again we allow the operation exchange col(i, j) in our A manipulations and continue trying to bring A to upper triangular form, as far as is possible; but, now we must reinterpret the effect of the column operation. Applying the operation exchange col(i, j) on A is equivalent to postmultiplying A by a certain matrix Q, equal to the identity matrix I, except that columns i and j have been exchanged. The matrix Q is its own inverse, that is, $QQ = I$. We have $A\mathbf{x} = AQQ\mathbf{x} = AQ\mathbf{x}'$, where $\mathbf{x}' = Q\mathbf{x}$, so if we perform an operation exchange col(i, j) on A, we also need to replace **x** by $\mathbf{x}' = Q\mathbf{x}$, that is, by a vector of unknowns with components x_i and x_j exchanged. We can think of the vector $\mathbf{x}'$ as a vector with components x_i' that equal x_{p_i}. Here $p_1, \ldots, p_n$ is a stored array defining some vector **p** with components that always are a permutation of the integers $1, 2, \ldots, n$. Initially, the p_i elements equal the integers $1, 2, \ldots, n$ in natural order. Each time we execute the operation exchange col(i, j) on A, we execute the operation exchange row(i, j) on **p** to define $\mathbf{x}'$. Our stored **p** vector gives us the connection between $\mathbf{x}'$ and **x**.

The final matrix A' that we attain is identical to the matrix 7.7, having a certain submatrix of elements $\doteq 0$, except that here because A' is square, the submatrix is square too. We can solve our problem if the A' submatrix is 1-square. The lone submatrix element a_{nn}' must equal 0. This is because the other diagonal elements are nonzero, yet det A', which equals the product of all diagonal elements, must be zero, because the matrix operations used never change the magnitude of det A'. To solve the system $A'\mathbf{x}' = \mathbf{0}$, we set x_n' equal to 1 and solve for the other components of $\mathbf{x}'$ by back substitution:

$$x'_n = 1$$

$$x'_{n-1} = -\frac{1}{a'_{n-1,n-1}} (a'_{n-1,n})$$

$$x'_{n-2} = -\frac{1}{a'_{n-2,n-2}} (a'_{n-2,n-1} x'_{n-1} + a'_{n-2,n})$$

$$\vdots \qquad\qquad \vdots$$

$$x'_1 = -\frac{1}{a'_{11}} (a'_{12} x'_2 + \cdots + a'_{1,n-1} x_{n-1} + a'_{1,n})$$

We obtain a vector of length 1 by multiplying all components of $\mathbf{x}'$ by $1/\text{len }\mathbf{x}'$, and we unscramble the effect of the column exchanges by consulting the $\mathbf{p}$ vector, taking x'_i as identical to x_{p_i}.

If the A' submatrix of elements $\doteq 0$ is not 1-square, then there is a difficulty. This time we cannot be certain that all submatrix elements are actually 0. A recomputation of A' at a higher precision might result in a smaller submatrix, but then again, it might not. Let us suppose for a moment that all submatrix elements are 0. We can obtain $n - r$ linearly independent solution vectors $\mathbf{x}_1, \ldots, \mathbf{x}_{n-r}$ by setting the variables $x'_{r+1}, x'_{r+2}, \ldots, x'_n$ equal successively to

$$
\begin{array}{cccc}
(1, & 0, & \ldots, & 0) \\
(0, & 1, & \ldots, & 0) \\
\vdots & \vdots & & \vdots \\
(0, & 0, & \ldots, & 1)
\end{array}
$$

each time solving for the rest of the variables x'_i by back substitution, adjusting values afterward to achieve a length of 1, and then reassigning variables by setting $x'_i = x_{p_i}$. This is in accordance with a theorem of linear algebra, namely

THEOREM 7.2: *If the* n*-square real (complex) matrix* A *has rank* r, *then there exist* n $-$ r *linearly independent solution vectors* $\mathbf{x}_1, \ldots, \mathbf{x}_{n-r}$ *to the matrix equation* Ax $=$ **0**, *and a vector* **x** *is a solution vector if and only if it can be written in the form*

$$\mathbf{x} = c_1 \mathbf{x}_1 + c_2 \mathbf{x}_2 + \cdots + c_{n-r} \mathbf{x}_{n-r}$$

where $c_1, c_2, \ldots, c_{n-r}$ *are constants.*

When there are two or more rows to the submatrix, it is risky to assume that all of its elements are zero. We cannot be confident in the solution vectors $\mathbf{x}_i$ unless the ambiguity of the submatrix elements is resolved. For the special case where A is a diagonal matrix, no computation at all is required, and solving our problem becomes purely a matter of determining which diagonal elements are 0. It is clear then that with this problem we cannot avoid encountering the nonsolvable problem of deciding whether two numbers are equal.

NONSOLVABLE PROBLEM 7.5: Given an n-square real (complex) matrix A, $n > 1$, with det $A = 0$, find to k decimals the elements of an n-vector $\mathbf{x}$ of length 1 such that $A\mathbf{x} = \mathbf{0}$.

If we know in advance what the rank of A is, then the difficulty in interpreting A' disappears. If the matrix A' we obtain is such that r equals the known rank, then all submatrix elements must be zero, and we find $n - r$ solution vectors of length 1 by the method described. And if r is less than the known rank, then A' must be recomputed at a higher precision.

SOLVABLE PROBLEM 7.6: Given an n-square real (complex) matrix A of known rank $r < n$, find to k decimals the elements of $n - r$ linearly independent n-vectors $\mathbf{x}_1$, $\mathbf{x}_2$, . . . , $\mathbf{x}_{n-r}$, all of length 1, such that $A\mathbf{x}_i = \mathbf{0}$ for all i.

If we know nothing about the n-square matrix A except that the rank r is less than n, that is, det $A = 0$, then we obtain a solvable problem only by reducing our requirements:

SOLVABLE PROBLEM 7.7: Given an n-square real (complex) matrix A, $n > 1$, with det A equal to zero, find to k decimals the elements of an n-vector $\mathbf{x}$ of length 1 such that $A\mathbf{x} = \mathbf{0}$, or else find to k decimals the elements of two or more n-vectors $\mathbf{x}_1$, $\mathbf{x}_2$, . . . , $\mathbf{x}_q$, all of length 1, such that len$(A\mathbf{x}_i) < 10^{-k}$ for all i.

The procedure to be followed here depends on the form of A'. If we find only a 1-square submatrix of zero approximants, then, as described previously, we obtain a vector $\mathbf{x}$ of length 1 satisfying the equation $A\mathbf{x} = \mathbf{0}$. If there is more than one row to the submatrix, then we assume that all submatrix elements are zero to get the vectors $\mathbf{x}_1$, $\mathbf{x}_2$, . . . , $\mathbf{x}_{n-r}$. These vectors must be tested with the original matrix A to determine whether len$(A\mathbf{x}_i) < 10^{-k}$ for all i. If this result is not obtained, and also if the elements of $\mathbf{x}$ or $\mathbf{x}_i$ are not found to k decimals, then we must recompute A' at an appropriate higher precision.

7.9. EIGENVALUES AND EIGENVECTORS

In many contexts the problem arises of finding for a square matrix A a nonzero vector $\mathbf{x}$ such that the matrix equation

$$A\mathbf{x} = \lambda \mathbf{x}$$

holds for some number λ, real or complex. The values possible for λ are called *eigenvalues*, and the vector **x** is called an *eigenvector*. We can rewrite the matrix equation as

$$(A - \lambda I)\mathbf{x} = 0$$

and in this form it is clear that there can be a nonzero vector **x** only if det $(A - \lambda I) = 0$. The matrix $A - \lambda I$ has the form

$$\begin{bmatrix} a_{11} - \lambda & a_{12} & \cdots & a_{1n} \\ a_{21} & a_{22} - \lambda & \cdots & a_{2n} \\ \vdots & \vdots & \ddots & \vdots \\ a_{n1} & a_{n2} & \cdots & a_{nn} - \lambda \end{bmatrix}$$

and $\det(A - \lambda I)$ is a polynomial in λ:

$$\det(A - \lambda I) = (a_{11} - \lambda)(a_{22} - \lambda) \cdots (a_{nn} - \lambda) + \text{other determinant terms}$$

$$\det(A - \lambda I) = (-1)^n(\lambda^n + c_{n-1}\lambda^{n-1} + \cdots + c_1\lambda + c_0) \tag{7.9}$$

The polynomial $\lambda^n + c_{n-1}\lambda^{n-1} + \ldots + c_1\lambda + c_0$ is called the *characteristic polynomial* of **A**. There are n eigenvalues $\lambda_1, \lambda_2, \ldots, \lambda_n$, corresponding to the n zeros of the characteristic polynomial.

There is no difficulty in accurately computing the coefficients of the characteristic polynomial, and in the next section practical methods of doing this are presented. By results of Chapter 6, we have

SOLVABLE PROBLEM 7.8: Find to k decimals the n eigenvalues of a real (complex) n-square matrix A.

Consider next the problem of determining for each eigenvalue λ_i an associated eigenvector $\mathbf{x}_i$. Solving the matrix equation

$$(A - \lambda_i I)\mathbf{x}_i = 0$$

for $\mathbf{x}_i$ is equivalent to solving a system of linear homogeneous equations for the components of $\mathbf{x}_i$, a problem considered in the preceding section. It is helpful here to introduce the concept of *similarity*. One n-square matrix B is similar to another n-square matrix A, if there is an n-square matrix P with an inverse P^{-1} such that

$$B = P^{-1}AP \tag{7.10}$$

The matrix P is said to *transform* A into B. From Equation 7.10 it follows that

$$PBP^{-1} = A$$

Thus P^{-1} transforms B into A, so the relation of similarity is symmetric. Similar matrices have the same characteristic polynomial, because

$$\det(B - \lambda I) = \det(P^{-1}AP - \lambda I) = \det(P^{-1}(A - \lambda I)P)$$

$$= \det P^{-1}\det(A - \lambda I)\det P = \det P^{-1}\det P \det(A - \lambda I)$$

$$= \det(P^{-1}P)\det(A - \lambda I) = \det I \det(A - \lambda I)$$

$$= \det(A - \lambda I)$$

Thus similar matrices A and B have identical eigenvalues. Moreover, if $\mathbf{x}_i$ is a λ_i eigenvector of B, then $\mathbf{y}_i = P\mathbf{x}_i$ is a λ_i eigenvector of A, since from $B\mathbf{x}_i = \lambda_i\mathbf{x}_i$ it follows that

$$BP^{-1}P\mathbf{x}_i = \lambda_i\mathbf{x}_i$$

$$PBP^{-1}P\mathbf{x}_i = \lambda_iP\mathbf{x}_i$$

$$A(P\mathbf{x}_i) = \lambda_i(P\mathbf{x}_i)$$

$$A\mathbf{y}_i = \lambda_i\mathbf{y}_i$$

Thus if we can find a matrix similar to A for which it is easy to find eigenvectors, then we also obtain eigenvectors for A by using the transformation matrix P.

Among the matrices similar to A, there exists a certain upper triangular matrix J, called the *Jordan normal form* of A, or more simply, the *Jordan form* of A, and defined in the theorem that follows. Let $J_r(\lambda)$ be the r-square matrix

$$J_r(\lambda) = \begin{bmatrix} \lambda & 1 & & & \\ & \lambda & 1 & & \\ & & \ddots & \ddots & \\ & & & \lambda & 1 \\ & & & & \lambda \end{bmatrix}_{r \times r} \tag{7.11}$$

The matrix $J_r(\lambda)$ is called a *Jordan block*. Its diagonal elements all equal λ, and all its elements just above the diagonal equal 1.

THEOREM 7.3: *For any* n-*square complex matrix* A *there exists an* n-*square matrix* P *with inverse* P^{-1} *such that* P^{-1} AP $=$ J *where* J, *the Jordan form of* A, *is an* n-*square matrix composed of Jordan blocks in the configuration*

$$J = \begin{bmatrix} J_{m_1}(\lambda_1) & & & \\ & J_{m_2}(\lambda_2) & & \\ & & \ddots & \\ & & & J_{m_t}(\lambda_t) \end{bmatrix} \tag{7.12}$$

The Jordan form of A is unique up to rearrangement of the Jordan blocks. These blocks are all in diagonal position, that is, their diagonal elements are also diagonal elements of J.

The block eigenvalues $\lambda_1, \lambda_2, \ldots, \lambda_t$ are eigenvalues of J and consequently of A too, but these block eigenvalues are not necessarily all distinct. If λ_j is an eigenvalue of A with multiplicity m, then λ_j must appear m times as a diagonal element of J and may be the diagonal element of more than one Jordan block. For example, if A is a 7-square matrix with eigenvalues λ_1, λ_2, and λ_3, with multiplicities 2, 2, and 3, respectively, then it is possible for J to take the form

$$\begin{bmatrix} \lambda_1 & 1 & & & & & \\ & \lambda_1 & & & & & \\ & & \lambda_2 & & & & \\ & & & \lambda_2 & & & \\ & & & & \lambda_3 & 1 & \\ & & & & & \lambda_3 & 1 \\ & & & & & & \lambda_3 \end{bmatrix}$$

with diagonal Jordan blocks $J_2(\lambda_1)$, $J_1(\lambda_2)$, $J_1(\lambda_2)$, and $J_3(\lambda_3)$.

It is sometimes difficult to actually compute the Jordan form. For instance, suppose the A matrix is

$$\begin{bmatrix} a & b \\ 0 & a \end{bmatrix}$$

and $b \doteq 0$. The Jordan form is

$$\begin{bmatrix} a & 0 \\ 0 & a \end{bmatrix} \quad \text{or} \quad \begin{bmatrix} a & 1 \\ 0 & a \end{bmatrix}$$

according as b is or is not equal to zero. (If $b \neq 0$, the transforming matrix P is a diagonal matrix with the diagonal elements 1, b^{-1}.) We see that again we encounter the nonsolvable problem of determining whether a number is zero.

NONSOLVABLE PROBLEM 7.9: For a real (complex) n-square matrix A, with $n > 1$, determine to k decimals the elements of the Jordan form of A.

It is not the diagonal elements of J that are difficult to determine, because these are eigenvalues and can be found to any number of correct decimals, but the 1's and 0's above the diagonal. The Jordan form, though sometimes difficult to compute, is helpful in understanding eigenvalue–eigenvector problems.

It is simple to determine the eigenvectors of a Jordan form J. The eigenvector equation for a Jordan block $J_r(\lambda)$ is $(J_r(\lambda) - \lambda I)\mathbf{x} = 0$, where the vector $\mathbf{x}$ has r components. Its solution is easily seen to be nonzero multiples of

$$\mathbf{x}^{(r)} = \begin{bmatrix} 1 \\ 0 \\ \vdots \\ 0 \end{bmatrix}$$

For each Jordan block $J_r(\lambda)$ appearing in J we obtain an eigenvector $\mathbf{x}$ by adding zero components to $\mathbf{x}^{(r)}$ in the obvious way to make an n-vector with a single 1 component. If there are t Jordan blocks in the J matrix, we obtain in this way t linearly independent eigenvectors $\mathbf{x}_1, \ldots, \mathbf{x}_t$, one vector associated with each block. If λ_o is a particular eigenvalue of J, it is easy to see that the equation $(J - \lambda_o I)\mathbf{x} = \mathbf{0}$ implies that any eigenvector associated with λ_o must be a linear combination of those eigenvectors associated with blocks $J_r(\lambda)$ with λ equal to λ_o.

When we compute eigenvalues to a certain number of correct decimal places, we can assign apparent multiplicities to them. An apparent multiplicity of 1 is always the correct multiplicity and the eigenvalue is *simple*. Suppose λ_o is a simple eigenvalue of A. Then λ_o is also a simple eigenvalue of J, and this implies that the matrix $J - \lambda_o I$ has rank $n - 1$. The rank of the matrix $A - \lambda_o I$ also is $n - 1$, for we have

$$A - \lambda_o I = PJP^{-1} - \lambda_o I = P(J - \lambda_o I)P^{-1}$$

The standard result below implies that the ranks of $A - \lambda_o I$ and $J - \lambda_o I$ are the same.

THEOREM 7.4: *Let* B *be any matrix of size* m $\times$ n. *If* C_1 *is an* m-*square matrix with det* $C_1 \neq 0$, *and* C_2 *is an* n-*square matrix with det* $C_2 \neq 0$, *then* B, C_1B, BC_2, *and* C_1BC_2 *all have the same rank.*

Solvable Problem 7.6 now implies

SOLVABLE PROBLEM 7.10: For a simple eigenvalue λ_o of a real (complex) n-square matrix A, find to k decimals the elements of an associated eigenvector $\mathbf{x}$ of length 1.

Suppose now that λ_o is an apparent multiple eigenvalue of A. Here we are uncertain whether our computed multiplicity is correct. Examining the Jordan form, we see that if there are q Jordan blocks associated with λ_o, then the rank of $J - \lambda_o I$ is $n - q$, and there are q linearly independent eigenvectors associated with λ_o. The ranks of $A - \lambda_o I$ and $J - \lambda_o I$ are the same; and if we were certain what this rank was, according to Solvable Problem 7.6, we could determine the correct number of linearly independent A eigenvectors for λ_o. But we can not be certain of a computed rank unless it turns out to equal $n - 1$. Solvable Problem 7.7 is helpful in this situation.

SOLVABLE PROBLEM 7.11: For an apparent multiple eigenvalue λ_o of a real (complex) n-square matrix A, find to k decimals the elements of a single associated eigenvector $\mathbf{x}$ of length 1, or else find to k decimals the elements of two or more vectors $\mathbf{x}_1, \ldots, \mathbf{x}_q$, each of length 1 and satisfying the inequality len($A\mathbf{x}_i - \lambda_o\mathbf{x}_i$) $< 10^{-k}$ for all i.

The vectors $\mathbf{x}_1, \ldots, \mathbf{x}_q$ may be called *near eigenvectors*.

We now turn to practical methods for handling the solvable problems of this section.

7.10. FINDING EIGENVALUES BY COMPUTING THE CHARACTERISTIC POLYNOMIAL

A procedure called Krylov's method can be used to obtain the characteristic polynomial of an n-square matrix A. Suppose the characteristic polynomial of A is

$$\lambda^n + c_{n-1}\lambda^{n-1} + \cdots + c_1\lambda + c_0$$

It is a consequence of Theorem 7.3, concerning the Jordan form J of A, that the matrix A satisfies the equation

$$A^n + c_{n-1}A^{n-1} + \cdots + c_1A + c_0I = O \qquad (7.13)$$

That is, if in the characteristic polynomial of A, we replace λ by A and interpret the result as a polynomial in A, substituting c_0I for the constant c_0, we obtain the zero matrix. This can be shown as follows: We have

$$z^n + c_{n-1}z^{n-1} + \cdots + c_1z + c_0 = (z - \lambda_1)^{m_1}(z - \lambda_2)^{m_2}\cdots(z - \lambda_t)^{m_t}$$

Let $J_r(\lambda_i)$ be a Jordan block of J associated with the eigenvalue λ_i. Then

$$(J_r(\lambda_i) - \lambda_iI)^r = O$$

where the identity matrix I is r-square, and $r \leq m_i$. This implies that

$$(J - \lambda_iI)^{m_i}$$

has a zero submatrix wherever J has a Jordan block associated with λ_i and nonzero elements only in the other block locations. As a consequence, we have

$$(J - \lambda_1I)^{m_1}(J - \lambda_2I)^{m_2}\cdots(J - \lambda_mI)^{m_t} = O$$

which leads to

$$J^n + c_{n-1}J^{n-1} + \ldots + c_iJ + c_0I = O$$

After we replace J by $P^{-1}AP$ in the matrix equation above, simplify, and then premultiply by P and postmultiply by P^{-1}, we obtain 7.13.

Equation 7.13 can be used to find the characteristic polynomial. Suppose $\mathbf{b}_0$ is an arbitrary n-vector. If we multiply Equation 7.13 on the right by $\mathbf{b}_0$ and set $\mathbf{b}_i$ equal to the vector $A^i\mathbf{b}_0$, then we obtain the equation

$$\mathbf{b}_n + c_{n-1}\mathbf{b}_{n-1} + \ldots + c_1\mathbf{b}_1 + c_0\mathbf{b}_0 = \mathbf{0}$$

or

$$c_0\mathbf{b}_0 + c_1\mathbf{b}_1 + \ldots + c_{n-1}\mathbf{b}_{n-1} = -\mathbf{b}_n \qquad (7.14)$$

The last equation is equivalent to n linear equations in the n unknowns c_0, $c_1, \ldots, c_{n-1}$. The initial vector $\mathbf{b}_0$ may be chosen as some convenient vector, for example,

$$\begin{bmatrix} 1 \\ 0 \\ 0 \\ \vdots \\ 0 \end{bmatrix} \quad \text{or} \quad \begin{bmatrix} 1 \\ 1 \\ 1 \\ \vdots \\ 1 \end{bmatrix}$$

and then the remaining vectors $\mathbf{b}_i$ are easily calculated according to the recurrence relation $\mathbf{b}_i = A\mathbf{b}_{i-1}$. Equation 7.14 can be put into the more conventional matrix form $B\mathbf{c} = -\mathbf{b}_n$, if we take B to be an n-square matrix with columns equal to the vectors $\mathbf{b}_0, \mathbf{b}_1, \ldots, \mathbf{b}_{n-1}$ and take $\mathbf{c}$ to be a vector with components $c_0, \ldots, c_{n-1}$. Thus with Krylov's method, we need only construct an n-square coefficient matrix B and vector $-\mathbf{b}_n$, using a simple recurrence relation, and then solve the corresponding linear equations to obtain the coefficients of the characteristic polynomial.

It is possible that we are unable to solve the equations, the coefficient matrix having a determinant $\doteq 0$. We may try repeating the Krylov procedure a few times with different initial vectors $\mathbf{b}_0$, but the difficulty may persist, as it would, for instance, if A were O. Thus with Krylov's method we need an alternate procedure. A simple backup procedure is to evaluate n determinants $\det(A - \lambda I)$ with λ chosen equal to n convenient numbers $a_1, a_2, \ldots, a_n$. For instance, we could set a_i equal to the integer i. If we set d_i equal to $\det(A - a_iI)$, then from Equation 7.9 defining the characteristic polynomial, we have

$$c_0 + c_1a_i + \ldots + c_{n-1}a_i^{n-1} = (-1)^n d_i - a_i^n \qquad i = 1, 2, \ldots, n$$

These n equations in c_i have the coefficient matrix

$$V(a_1, a_2 \ldots, a_n) = \begin{bmatrix} 1 & a_1 & a_1^2 & \cdots & a_1^{n-1} \\ 1 & a_2 & a_2^2 & \cdots & a_2^{n-1} \\ \vdots & \vdots & \vdots & \vdots\vdots\vdots & \vdots \\ 1 & a_n & a_n^2 & \cdots & a_n^{n-1} \end{bmatrix} \qquad (7.15)$$

A matrix of this form is called a Vandermonde matrix.

We show in the last paragraph of this section that our Vandermonde matrix has a nonzero determinant, so success in determining the coefficients c_i is to be expected, as long as the precision of computation is sufficiently high.

The coefficients of the characteristic polynomial of a problem matrix A, are obtained as correctly ranged numbers when computed by either of the two suggested methods. The eigenvalues of this polynomial, when they are determined by the methods of Chapter 6, are also correctly ranged, and an apparent multiplicity for each will also be known. For each simple eigenvalue, we can determine the associated eigenvector's components, and for each apparently multiple eigenvalue, we can determine either the associated eigenvector's components or the components of a set of *near eigenvectors* by employing the methods described earlier. Of course doing this for each eigenvalue individually is inefficient, and a better method for computing eigenvectors is described in the next section.

It is easy to show by an indirect argument that det $V(a_1, \ldots, a_n)$ is nonzero if all the numbers a_i are distinct. For an n-square matrix A, if there are no nonzero n-vectors $\mathbf{x}$ satisfying the equation $A\mathbf{x} = \mathbf{0}$, then this implies det $A \neq 0$. Suppose then that all the numbers a_i are distinct and that the n-vector $\mathbf{d}$ satisfies the equation

$$V(a_1, \ldots, a_n)\mathbf{d} = \mathbf{0} \qquad (7.16)$$

The components d_i of $\mathbf{d}$ define a polynomial $P(z)$, given by

$$P(z) = d_1 + d_2 z + d_3 z^2 + \cdots d_n z^{n-1}$$

The j-th component of Equation 7.16 implies that $P(a_j) = 0$, that is, a_j is a zero of the polymonial $P(z)$. But then $P(z)$ has n distinct zeros yet its degree can be at most $n - 1$. We must conclude that $\mathbf{d} = \mathbf{0}$, and this proves det $V(a_1, \ldots, a_n) \neq 0$.

7.11. FINDING EIGENVALUES AND EIGENVECTORS BY DANILEVSKY'S METHOD

Before we can present Danilevsky's method, we need to become familiar with the properties of *companion* matrices. An n-square matrix C is a *companion matrix* if it has the form

$$C = \begin{bmatrix} 0 & 0 & \cdots & 0 & -c_0 \\ 1 & 0 & \cdots & 0 & -c_1 \\ \vdots & \vdots & \vdots\vdots\vdots & \vdots & \vdots \\ 0 & 0 & \cdots & 1 & -c_{n-1} \end{bmatrix}$$

A companion matrix has all elements zero except for the elements just below the diagonal, which equal 1, and the elements in the last column. Its characteristic polynomial is

$$\det(C - \lambda I) = (-1)^n(\lambda^n + c_{n-1}\lambda^{n-1} + \cdots + c_1\lambda + c_0) \qquad (7.17)$$

To see this, compute the determinant of

$$C - \lambda I = \begin{bmatrix} -\lambda & 0 & \cdots & 0 & -c_0 \\ 1 & -\lambda & \cdots & 0 & -c_1 \\ \vdots & \vdots & \vdots\vdots\vdots & \vdots & \vdots \\ 0 & 0 & \cdots & 1 & -\lambda - c_{n-1} \end{bmatrix}$$

by a cofactor expansion along the last column. The cofactor of the element $-c_0$ is $(-1)^{n+1}$ times the determinant of an upper triangular matrix with all diagonal elements equal to 1. The cofactor of the element $-\lambda - c_{n-1}$ is $(-1)^{n+n} = +1$ times the determinant of a lower triangular matrix with all diagonal elements equal to $-\lambda$. This verifies that the coefficients shown in Equation 7.17 are correct for the leading two terms of the polynomial and the constant term. The cofactor of any other element in the last column, say $-c_j$, is $(-1)^{n+j+1}$ times the determinant of a matrix with the structure

$$\begin{bmatrix} B_1 & O \\ O & B_2 \end{bmatrix}$$

The j-square matrix B_1 is lower triangular with diagonal elements equal to $-\lambda$, and the $(n - 1 - j)$-square matrix B_2 is upper triangular with diagonal elements equal to 1. This leads to a contribution to $\det(C - \lambda I)$ of $(-c_j)(-1)^{n+j+1}(-\lambda)^j = (-1)^n c_j \lambda^j$, in agreement with Equation 7.17.

For a companion matrix it is easy to compute eigenvalues, because the elements of its last column give the coefficients of its characteristic polynomial. If $\lambda_1, \lambda_2, \ldots, \lambda_n$ are the eigenvalues, and these are all distinct, then for the Vandermonde matrix V equal to $V(\lambda_1, \lambda_2 \ldots, \lambda_n)$, we find that the matrix equation $VC = JV$ is true where J is the diagonal matrix

$$\begin{bmatrix} \lambda_1 & 0 & \cdots & 0 \\ 0 & \lambda_2 & \cdots & 0 \\ \vdots & \vdots & \ddots & \vdots \\ 0 & 0 & \cdots & \lambda_n \end{bmatrix}$$

Thus we have the matrix equation $VCV^{-1} = J$, so we obtain the Jordan form of

C by transforming C with V^{-1}. If an eigenvalue λ_i has a multiplicity m greater than 1, the matrix V must be changed. A block of m rows in V must be assigned to λ_i with the structure

$$
\begin{bmatrix}
0 & 0 & 0 & 0 & \cdots & 1 & \binom{m}{m-1}\lambda_i & \cdots & \binom{n-1}{m-1}\lambda_i^{n-m} \\
\vdots & \vdots & \vdots & \vdots & \vdots\vdots\vdots & \vdots & \vdots & \vdots\vdots\vdots & \vdots \\
0 & 0 & 1 & \binom{3}{2}\lambda_i & \cdots & \binom{m-1}{2}\lambda_i^{m-3} & \binom{m}{2}\lambda_i^{m-2} & \cdots & \binom{n-1}{2}\lambda_i^{n-3} \\
0 & 1 & \binom{2}{1}\lambda_i & \binom{3}{1}\lambda_i^2 & \cdots & \binom{m-1}{1}\lambda_i^{m-2} & \binom{m}{1}\lambda_i^{m-1} & \cdots & \binom{n-1}{1}\lambda_i^{n-2} \\
1 & \lambda_i & \lambda_i^2 & \lambda_i^3 & \cdots & \lambda_i^{m-1} & \lambda_i^m & \cdots & \lambda_i^{n-1}
\end{bmatrix}
$$

If we let V_m denote the above $m \times n$ matrix, then we find that the matrix equation $V_m C = J_m(\lambda_i)V_m$ is true, where $J_m(\lambda_i)$ is an m-square Jordan block. To see this, we need two results. One is the binomial identity

$$
\binom{s}{j} = \binom{s-1}{j} + \binom{s-1}{j-1}
$$

which is valid when the integer s is larger than the positive integer j. Also, not only does λ_i satisfy the polynomial equation

$$
-(c_0 + c_1\lambda + c_2\lambda^2 + \cdots + c_{n-1}\lambda^{n-1}) = \lambda^n
$$

but because it has multiplicity m, it satisfies the j-th derivative of this equation with respect to λ, for j between 1 and $m - 1$. If we take the j-th derivative and then divide by $j!$, we obtain the equation

$$
-\left(c_j + c_{j+1}\binom{j+1}{j}\lambda + \cdots + c_{n-1}\binom{n-1}{j}\lambda^{n-1-j}\right) = \binom{n}{j}\lambda^{n-j}
$$

Thus if the characteristic equation of a companion matrix has eigenvalues $\lambda_1, \lambda_2, \ldots, \lambda_t$ with respective multiplicities $m_1, m_2, \ldots, m_t$, and we form a matrix V composed of blocks as shown above for the distinct eigenvalues, we have the matrix equation $VC = JV$ where J is the Jordan form for C. The argument of the last paragraph of the preceding section can be repeated for the changed Vandermonde matrix V to show that it too must have a nonzero determinant. Thus V has an inverse V^{-1}, so we have $VCV^{-1} = J$, and C is transformed to Jordan form by V^{-1}. The Jordan form of a companion matrix is somewhat restricted, in that an eigenvalue λ_i can have only one associated Jordan block.

Given an n-square problem matrix A, Danilevsky's method consists of attempting to find a matrix Q such that $Q^{-1}AQ$ is a companion matrix C. If we are successful in finding Q and C, then the last column of C gives us the coefficients of the common characteristic polynomial of C and A. By methods of Chapter 6, we obtain the eigenvalues as the distinct zeros of this polynomial, each with an assigned apparent multiplicity. This allows us to construct the matrices V and V^{-1} and to transform C into Jordan form J. Since $Q^{-1}AQ = C$ and $VCV^{-1} = J$, if we set P equal to QV^{-1}, then $P^{-1}AP = J$. The eigenvector of J associated with an eigenvalue λ_i is a vector $\mathbf{x}_i$ with all zero components except for a single 1 component aligned with the starting row of the Jordan block associated with λ_i. The corresponding eigenvector $\mathbf{y}_i$ of $\mathbf{A}$ equals $P\mathbf{x}_i$, as was shown in Section 7.9. Thus the eigenvectors of A are certain column vectors of P.

The matrices Q and C are computed in stages. That is, a series of transformations are made, gradually bringing A to companion form. Suppose the matrix A has been transformed by a matrix Q to a form A_i, which has all columns to the left of column i in the required form, that is,

$$
A_i = \begin{bmatrix}
0 & & & & a_{1i} & \cdots & a_{1n} \\
1 & & & & a_{2i} & \cdots & a_{2n} \\
& 1 & & & a_{3i} & \cdots & a_{3n} \\
& & \ddots & & \vdots & \vdots\vdots\vdots & \vdots \\
& & & 1 & a_{ii} & \cdots & a_{in} \\
& & & & a_{i+1,i} & \cdots & a_{i+1,n} \\
& & & & \vdots & \vdots\vdots\vdots & \vdots \\
& & & & a_{ni} & \cdots & a_{nn}
\end{bmatrix}
$$

The elements a_{ij} shown above change as A_i is transformed into A_{i+1}. At the beginning of operations $A = A_1$, $Q = I$, and no columns are in correct form. The process ends when the transform A_n is obtained, the elements in the last column being interpreted now as defining the characteristic polynomial of a companion matrix.

We want to transform A_i so that in column i we have the needed structure of all 0s except for a single 1 below the diagonal position. We require an n-square matrix Q_i such that $Q_i^{-1} A_i Q_i = A_{i+1}$. Each time we perform any transformation, the old Q is postmultiplied by the transforming matrix, Q_i in this case, to become the new current Q matrix. If $a_{i+1,i} \neq 0$, the needed matrix Q_i is equal to the identity matrix I except that column $i + 1$ is replaced by column i of A_i. This matrix is shown below

$$
Q_i = \begin{bmatrix}
1 & & & & a_{1i} & & & \\
& 1 & & & a_{2i} & & & \\
& & \ddots & & \vdots & & & \\
& & & 1 & a_{ii} & & & \\
& & & & a_{i+1,i} & & & \\
& & & & a_{i+2,i} & 1 & & \\
& & & & \vdots & & \ddots & \\
& & & & a_{ni} & & & 1
\end{bmatrix}
\quad \text{col } i + 1
$$

The inverse of this matrix is

$$
Q_i^{-1} = \begin{bmatrix}
1 & & & & -\dfrac{a_{1i}}{a_{i+1,i}} & & & \\
 & 1 & & & -\dfrac{a_{2i}}{a_{i+1,i}} & & & \\
 & & \ddots & & \vdots & & & \\
 & & & 1 & -\dfrac{a_{ii}}{a_{i+1,i}} & & & \\
 & & & & \dfrac{1}{a_{i+1,i}} & & & \\
 & & & & -\dfrac{a_{i+2,i}}{a_{i+1,i}} & 1 & & \\
 & & & & \vdots & & \ddots & \\
 & & & & -\dfrac{a_{ni}}{a_{i+1,i}} & & & 1
\end{bmatrix}
$$

col $i + 1$

Because the first i columns of Q_i are identical to corresponding columns of I, the first i columns of the product $A_i Q_i$ are the same as those of A_i. The premultiplication of $A_i Q_i$ by Q_i^{-1} completes the transform of A_i to a matrix of form A_{i+1}, with one more column in the correct companion matrix form.

If we find $a_{i+1,i} \doteq 0$, we cannot perform the transformation because Q_i^{-1} is not defined, and instead we examine the elements $a_{i+2,i}, a_{i+3,i}, \ldots, a_{ni}$. With the first element $\ne 0$, say a_{ji}, we perform the operation exchange row $(i + 1, j)$ to bring this element into the $a_{i+1,j}$ position without disturbing the companion matrix columns already obtained. This operation is equivalent to premultiplying A_i by a matrix I' identical to I except that rows $i + 1$ and j have been exchanged. The matrix I' is its own inverse, so to complete the transformation, the needed postmultiplication by I' is done by performing the operation exchange col$(i + 1, j)$. After these operations are performed and Q updated, we can transform our matrix to the form A_{i+1}.

It may happen when working with A_i that we find no element $\ne 0$ among the elements $a_{i+1,i}, a_{i+2,i}, \ldots, a_{in}$. In this case we give up trying to transform A_i to companion matrix form and try instead to transform it to the form

$$
\begin{bmatrix} C_1 & O \\ O & C_2 \end{bmatrix} \tag{7.18}
$$

Here there are two companion matrices of various sizes, both in diagonal position. This more general goal is sometimes needed with Danilevsky's method, because the Jordan form of a companion matrix is restricted, in that an eigenvalue of multiplicity greater than 1 can have only a single Jordan block. If the Jordan form of A has several Jordan blocks associated with a multiple eigenvalue, the more general goal becomes necessary.

When the A_i impasse is encountered, the structure of the matrix A_i may be interpreted as

$$\begin{bmatrix} C_1 & B \\ O & D \end{bmatrix} \tag{7.19}$$

The i-square submatrix occupying the first i rows and columns of A_i is in companion form, and the submatrix below it is a zero submatrix, or more exactly, a submatrix with all elements either $= 0$ (the first $i - 1$ columns) or $\doteq 0$ (column i). We allow the zero matrices of 7.18 to have elements of the two varieties. By a series of transformations, we convert rows $i, i. - 1, \ldots, 2$ of the submatrix B to rows of zeros, preserving the submatrix C_1 and the zero submatrix below it. After each transformation, the nonzero elements of B change, as do the elements of D, but to keep our notation simple, we do not bother to differentiate between the various B and D submatrices. Suppose we have brought A_i to a form $A^{(q)}$ where all rows of the changing matrix B below row q are zero. We can clear row q of B by a process similar to the one described earlier for clearing a column of A_i, but here we use the 1s of the matrix C_1 to do the clearing. Let S_q be a matrix identical to I except that row $q - 1$ has been replaced by row q of the B part of $A^{(q)}$. This matrix is shown below.

$$S_q = \begin{bmatrix} 1 & & & & & & & & \\ & \ddots & & & & & & & \\ & & 1 & 0 & \cdots & 0 & b_{q1} & \cdots & b_{q,n-i} \\ & & & 1 & & & & & \\ & & & & \ddots & & & & \\ & & & & & 1 & & & \\ & & & & & & 1 & & \\ & & & & & & & \ddots & \\ & & & & & & & & 1 \end{bmatrix} \text{row } q - 1$$

The matrix S_q^{-1} is identical to S_q except that all elements b_{qj} are replaced by $-b_{qj}$. To see the effect of transforming $A^{(q)}$ by S_q^{-1} to obtain $A^{(q-1)} = S_q A^{(q)} S_q^{-1}$, notice that $A^{(q)} S_q^{-1}$ has a cleared row q to the right of column i and that this is not disturbed by the premultiplication of S_q. The new submatrix B has zeros in row q, and the C_1 and O submatrices have not changed in form.

Transformations in succession by $S_i^{-1}, S_{i-1}^{-1}, \ldots, S_2^{-1}$ bring our matrix to a form where rows 2 through i of the B submatrix are all zeros. If the B elements in row 1 now $\doteq 0$, then the transformation to the general form 7.18 is complete, and the entire Danilevsky process can be applied to the submatrix D, which becomes our new problem matrix. However, when carrying out a transformation on this submatrix, we must be careful to extend the reach of any transformation on it beyond the D region to the regions above and to the left. This is because some elements there are approximate zeros rather than exact zeros, and we want their ranges to correctly reflect their computational error at all times.

Of course it may happen that there is an element $\neq 0$ in row 1 of the final B submatrix. Suppose we find $b_{1t} \neq 0$. This element is brought to a new position by transforming the problem matrix by a matrix U_t of the form shown below. The submatrix in the upper left of U_t is t-square.

$$
U_t = \begin{bmatrix}
0 & 1 & 0 & \cdots & 0 & & & \\
0 & 0 & 1 & \cdots & 0 & & & \\
\vdots & \vdots & \vdots & \ddots & \vdots & & & \\
0 & 0 & 0 & \cdots & 1 & & & \\
1 & 0 & 0 & \cdots & 0 & & & \\
& & & & & 1 & & \\
& & & & & & \ddots & \\
& & & & & & & 1
\end{bmatrix}
$$

Postmultiplication of our problem matrix by U_t brings column t to the column 1 position and moves columns 1 through $t - 1$ back one column. The inverse of U_t is its own transposition. Premultiplication now by U_t^{-1} performs a similar manipulation of rows 1 through t. After the transformation of our problem matrix is complete, its upper left submatrix in the first $i + 1$ rows and i columns has the form

$$
\begin{bmatrix}
0 & 0 & \cdots & 0 \\
b_{1t} & 0 & \cdots & 0 \\
0 & 1 & \cdots & 0 \\
\vdots & \vdots & \ddots & \vdots \\
0 & 0 & \cdots & 1
\end{bmatrix}_{(i+1)\times i}
$$

The transformation by U_t has possibly caused nonzero elements to appear in the first column below row $i + 1$. In any case the form of our problem matrix is now such that if we restart the Danilevsky method at column 1, we cannot again encounter the impasse that made us attempt to obtain multiple companion matrices until we are at least one column further than before.

With these changes in the Danilevsky procedure it is clear that we must eventually succeed in transforming the problem matrix to a form

$$
C = \begin{bmatrix}
C_1 & & & \\
& C_2 & & \\
& & \ddots & \\
& & & C_s
\end{bmatrix}
\tag{7.20}
$$

where all submatrices C_i are companion matrices. A frequent case is $s = 1$, there being only a single companion matrix. The elements of C that do not belong to any of the various companion submatrices are all zeros or approximate zeros. Because some of these elements may not be exact zeros, if we find correctly ranged eigenvalue approximations to the various companion submatrices, using their easily obtained characteristic polynomials, we cannot presume these eigenvalues are correctly ranged for C. Nevertheless for each companion matrix C_i, eigenvalue approximations can be found with computed apparent multiplicities, using the method of Section 6.5, and with this data a Vandermonde matrix V_i for C_i can be constructed. We form the matrix

$$V = \begin{bmatrix} V_1 & & & \\ & V_2 & & \\ & & \ddots & \\ & & & V_s \end{bmatrix}$$

to correspond to the structure 7.20. Using this matrix and its inverse V^{-1}, we can transform C to "approximate" Jordan form J'. Thus in all cases we can construct a matrix P equal to QV^{-1} such that $P^{-1}AP = J'$. The matrix P is used to obtain eigenvector approximations. We consider next how to bound the error of our approximate eigenvalues and approximate eigenvectors.

7.12. ERROR BOUNDS FOR DANILEVSKY'S METHOD

Danilevsky's method often delivers an end matrix $J,'$ which is close to a diagonal matrix. This occurs when the eigenvalue approximations are simple, their computed apparent multiplicities all found to equal 1. In this case, a result due to Gershgorin can be used to bound both eigenvalue error and eigenvector error.

THEOREM **7.5**: *Let B be any n-square matrix with real or complex elements. In the complex plane define the n disks*

$$|z - b_{ii}| \le r_i \qquad i = 1, 2, \ldots, n \tag{7.21}$$

where

$$r_i = \sum_{\substack{j=1 \\ j \ne i}}^{n} |b_{ij}|$$

These n disks may or may not overlap; in general they form s disjoint connected sets, where s = n only if the disks do not overlap. Every eigenvalue of B lies in one of these connected sets, and each connected set consisting of t disks contains t eigenvalues, counting multiplicities.

To prove the theorem, let λ be any eigenvalue of B, and let $\mathbf{x}$ be an associated eigenvector, adjusted by multiplication by a constant so that the component x_{i0} of largest magnitude equals 1. All other components of $\mathbf{x}$ then have magnitude ≤ 1. The matrix equation $\lambda \mathbf{x} = B\mathbf{x}$ may be converted to element equations:

$$(\lambda - b_{ii})x_i = \sum_{\substack{j=1 \\ j \neq i}}^{n} b_{ij}x_j \qquad i = 1, 2, \ldots, n \qquad (7.22)$$

Taking absolute values, we have

$$|\lambda - b_{ii}||x_i| = \left| \sum_{\substack{j=1 \\ j \neq i}}^{n} b_{ij}x_j \right| \leq \sum_{\substack{j=1 \\ j \neq i}}^{n} |b_{ij}||x_j| \leq \sum_{\substack{j=1 \\ j \neq i}}^{n} |b_{ij}| \cdot 1 = r_i \qquad i = 1, 2, \ldots, n$$

Because $x_{i0} = 1$, for equation i_0 we have

$$|\lambda - b_{i0,i0}| \leq r_{i0}$$

Thus λ lies in one of the disks, and therefore in one of the s connected sets.

To prove the rest of the theorem, consider the matrix

$$B(u) = (1 - u)J + uB$$

where J is a diagonal matrix with the same diagonal elements as B. As the real parameter u varies from 0 to 1, the matrix $B(u)$ varies from J to B. When $u = 0$, the disks become points and all parts of the theorem are correct. As u varies toward 1, the disk radii grow, and the eigenvalues move continuously in the complex plane yet cannot leave the disks. Consequently the number of disks in a connected set and the number of associated eigenvalues always match.

Suppose now the Danilevsky method has delivered an end matrix B that is near-diagonal. The matrix B may be represented as a sum of two parts, a diagonal matrix, which also is a Jordan form J and a discrepancy matrix E:

$$B = J + E = \begin{bmatrix} b_{11} & & & \\ & b_{22} & & \\ & & \ddots & \\ & & & b_{nn} \end{bmatrix} + \begin{bmatrix} 0 & b_{12} & \cdots & b_{1n} \\ b_{21} & 0 & \cdots & b_{2n} \\ \vdots & \vdots & \ddots & \vdots \\ b_{n1} & b_{n2} & \cdots & 0 \end{bmatrix}$$

The radius r_q equals the sum of the absolute values of the E matrix elements that are in row q, and bounds the error of the J diagonal element b_{qq} if the disk corresponding to this element is disjoint from the other disks. Let us assume this to be the case. Then we may set $\lambda_q = b_{qq} \oplus r_q$ to obtain a correctly ranged eigenvalue. Let $\mathbf{x}_q$ be the eigenvector associated with λ_q with its q-th component set equal to 1. The other components of $\mathbf{x}_q$ cannot have a magnitude larger than

1, otherwise, if x_{i_0} is the component of $\mathbf{x}_q$ that is largest in magnitude, repeating the steps of the proof of the preceding theorem, we obtain $|\lambda_q - b_{i_0,i_0}| \le r_{i_0}$, so λ_q lies in the b_{i_0,i_0} disk, contradicting our assumption that the b_{qq} disk was disjoint.

We can bound the magnitude of any component x_i of $\mathbf{x}_q$ different from x_q. Using Equation 7.22 for $i \ne q$, after taking absolute values, we have

$$|\lambda_q - b_{ii}||x_i| \le \sum_{\substack{j=1 \\ j \ne i}}^{n} |b_{ij}||x_k| \le \sum_{\substack{j=1 \\ j \ne i}}^{n} |b_{ij}| \cdot 1 = r_i$$

Hence

$$|x_i| \le \frac{r_i}{|\lambda_q - b_{ii}|} = \rho_i$$

and we may take x_i as $0 \oplus \rho_i$. Now that we have correctly ranged values for all the components of our eigenvector $\mathbf{x}_q$, we can form the A eigenvector $\mathbf{y}_q$ using the equation $\mathbf{y}_q = P\mathbf{x}_q$, and then convert this vector to a unit vector in the usual way.

Suppose now that the b_{qq} disk is not disjoint from the other disks, but is part of a disjoint composite formed by t disks. In this case we take λ_q as $b_{qq} \oplus R$ where $R = 2 \Sigma r_i$, the sum being over all indices i such that the b_{ii} disk is part of the composite. Now λ_q is taken as an eigenvalue of apparent multiplicity t. No longer do we attempt to find an eigenvalue for λ_q but try instead to supply t near eigenvectors using vectors $\mathbf{x}_i$ with all zero components except for a single 1 component in positions appropriate to the set of overlapping disks. The corresponding vectors $\mathbf{y}_i$ for the matrix A then are various columns of P, and these are converted to unit vectors in the usual way. Each of the final vectors $\mathbf{y}_i$ must be tested to determine whether the equation len $(A\mathbf{y}_i - \lambda_q\mathbf{y}_i) < 10^{-k}$ holds.

The eigenvalues and the components of the eigenvectors and near eigenvectors are tested to determine whether k correct decimals have been obtained, and if not, the Danilevsky computation is repeated at an appropriate higher precision.

Now we consider the case where the final matrix of Danilevsky's method is not a near diagonal matrix. Again we represent this final matrix B as a sum of two matrices, a Jordan form matrix J and a discrepancy matrix E. Clearly if an element of J or E is outside the Jordan block domains, then the element is 0 if it is in J and equals b_{ij} if it is in E. The plan for elements of J and E that are in a Jordan block domain, can be illustrated by showing just the elements of the first Jordan block. We presume this first block is m-square and show first the J elements and below them the E elements.

$$
\begin{bmatrix}
b_{11} & 1 & & & & \\
 & b_{11} & 1 & & & \\
 & & \ddots & \ddots & & \\
 & & & \ddots & \ddots & \\
 & & & & b_{11} & 1 \\
 & & & & & b_{11}
\end{bmatrix}
\tag{7.23}
$$

$$
\begin{bmatrix}
0 & b_{12} - 1 & b_{13} & \cdots & b_{1,m-1} & b_{1m} \\
b_{21} & b_{22} - b_{11} & b_{23} - 1 & \cdots & b_{2,m-1} & b_{2m} \\
\vdots & \vdots & \vdots & \vdots\vdots\vdots & \vdots & \vdots \\
\vdots & \vdots & \vdots & \vdots\vdots\vdots & \vdots & \vdots \\
b_{m-1,1} & b_{m-1,2} & b_{m-1,3} & \cdots & b_{m-1,m-1} - b_{11} & b_{m-1,m} - 1 \\
b_{m1} & b_{m2} & b_{m3} & \cdots & b_{m,m-1} & b_{mm} - b_{11}
\end{bmatrix}
$$

As before let r_i equal the sum of the absolute values of the matrix E elements in row i. This time, to keep our error bounds simple in form, we use only the quantity r defined by the equation $r = \max_j r_j$. Suppose $\mathbf{x}$ is a λ eigenvector of B with the component of largest magnitude being one of the components $x_1, \ldots, x_m$ associated with our first block. Suppose also that our eigenvector is multiplied by a constant to make the component of largest magnitude equal 1. The element equations are

$$
(\lambda - b_{11})x_1 = x_2 + \sum_{j=1}^{n} e_{1j}x_j
$$

$$
(\lambda - b_{11})x_2 = x_3 + \sum_{j=1}^{n} e_{2j}x_j
$$

$$
\vdots \qquad\qquad \vdots
$$

$$
(\lambda - b_{11})x_{m-1} = x_m + \sum_{j=1}^{n} e_{m-1,j}x_j
$$

$$
(\lambda - b_{11})x_m = \sum_{j=1}^{n} e_{mj}x_j
\tag{7.24}
$$

Here the quantities e_{ij} are the elements of the discrepancy matrix E. After we take absolute values, the first equation becomes

$$
|\lambda - b_{11}||x_1| \leq |x_2| + \sum_{j=1}^{n} |e_{1j}||x_j| \leq |x_2| + \sum_{j=1}^{n} |e_{1j}| \cdot 1 \leq |x_2| + r
$$

and, in general, we have

$$|\lambda - b_{11}||x_j| \le |x_{j+1}| + r \qquad j = 1, 2, \ldots, m - 1$$

$$|\lambda - b_{11}||x_m| \le r \qquad\qquad\qquad\qquad (7.25)$$

If the component that equals 1 is x_1, then after we multiply the first inequality above by $|\lambda - b_{11}|^{m-1}$, the second by $|\lambda - b_{11}|^{m-2}$, and so on down to the last equation, which is multiplied by 1, and then add all these inequalities and cancel any identical terms on both sides of the inequality, we obtain the relation

$$|\lambda - b_{11}|^m \le r|\lambda - b_{11}|^{m-1} + \ldots + r|\lambda - b_{11}| + r$$

or

$$|\lambda - b_{11}|^m - r|\lambda - b_{11}|^{m-1} - \ldots - r|\lambda - b_{11}| - r \le 0 \qquad (7.26)$$

The polynomial

$$z^m - rz^{m-1} - \ldots - rz - r$$

is positive if z is real and exceeds a bounding radius R on its zeros (see Theorem 6.2). So the inequality 7.26 implies $|\lambda - b_{11}| \le R$, and using the bounding radius given in the corollary to Theorem 6.2, we have

$$|\lambda - b_{11}| \le \max_{j=1}^m (mr)^{1/j} \qquad (7.27)$$

Note that $(mr)^{1/j}$ increases with increasing j if $mr < 1$. We will assume $mr < 1$ (we can always increase precision and start anew if we do not find this), so that we have the easily calculated bound

$$|\lambda - b_{11}| \le (mr)^{1/m} \qquad (7.28)$$

If we suppose that the largest modulus component of our eigenvector is not x_1 but some other component x_j associated with the first Jordan block, we obtain even sharper bounds on $|\lambda - b_{11}|$. Instead of the inequality 7.26, we obtain by a similar method the inequality

$$|\lambda - b_{11}|^{m-j+1} - r|\lambda - b_{11}|^{m-j} - \ldots - r|\lambda - b_{11}| - r \le 0$$

which leads to $|\lambda - b_{11}| \le ((m - j + 1)r)^{1/(m-j+1)}$. Thus the inequality 7.28 always can be used and defines a disk in the complex plane. We obtain a bounding disk for each Jordan block of B. Suppose a certain number of these disks overlap, forming a connected set, and let t equal the sum of the row sizes of the Jordan blocks corresponding to these disks. In the same way as was shown in the proof of the Gershgorin theorem, this set must contain t eigenvalues of B counting multiplicities.

The procedure to be followed when the matrix B approximates a nondiagonal Jordan form is similar to the procedure for the diagonal case. For the first Jordan block a correctly ranged eigenvalue is $\lambda_1 = b_{11} \oplus (mr)^{1/m}$ and similarly for the other blocks. If a disk for a Jordan block is discrete, we try to obtain eigenvector components for this block, and if the disk is part of a composite figure, we supply one near eigenvector for each block associated with the composite figure, these being various column vectors of the matrix P.

We consider the problem of determining an eigenvector for a Jordan block when the bounding disk is discrete. Again we assume the eigenvector has been multiplied by a constant so that its component of largest magnitude equals 1, and to illustrate the general procedure, we assume the eigenvector is associated with λ_1, the first block eigenvalue. By similar reasoning as before, the eigenvector's 1 component must be associated with the first block, that is, it must be one of the components $x_1, x_2, \ldots, x_m$. If the bound obtained by 7.27 is less than $\frac{1}{2}$, then it cannot be the case that the 1 component of our eigenvector is not x_1 but some other component x_j associated with the first Jordan block. If we assume this is the case, we arrive at a contradiction. The element Equation 7.24 for row $j - 1$ can be written

$$x_j = (\lambda - b_{11})x_{j-1} - \sum_{i=1}^{n} e_{j-1,i}x_i$$

After taking absolute values, setting $|x_j|$ equal to 1, and using the inequality $|x_i| \leq 1$ for all other components, we obtain the relation

$$1 \leq |\lambda - b_{11}| + r$$

which is impossible because both $|\lambda - b_{11}|$ and r are less than $\frac{1}{2}$.

Thus if r is small enough, the only eigenvectors possible for the matrix B are eigenvectors whose component of largest modulus, x_q, made equal to 1, is such that q equals the starting row index of a Jordan block. As in the diagonal B case we need to obtain correctly ranged values for the other components of such a vector. Continuing with the case of the λ_1 eigenvector, the first $m - 1$ equations of the set 7.24 can be written as

$$x_j = (\lambda_1 - b_{11})x_{j-1} - \sum_{i=1}^{n} e_{j-1,i}x_i \qquad j = 2, 3, \ldots, m$$

Taking absolute values, this leads to the inequalities

$$|x_j| \leq |\lambda_1 - b_{11}| + r \qquad j = 2, 3, \ldots, m$$

for the components other than x_1 that are "associated" with the first block, that is, have indices matching the row indices of the first block. For the components associated with some other block, starting in row s, which components we denote by $x_1', x_2', \ldots, x_{m'}'$, the block being m'-square, we obtain the component bounds below by taking absolute values of equations for this block analogous to the Equations 7.25.

$$|x_{m'}'| \leq \frac{r}{|\lambda_1 - b_{ss}|}$$

$$|x_{m'-1}'| \leq \frac{r}{|\lambda_1 - b_{ss}|} + \frac{r}{|\lambda_1 - b_{ss}|^2}$$

$$\vdots \qquad \vdots$$

$$|x_1'| \leq \frac{r}{|\lambda_1 - b_{ss}|} + \frac{r}{|\lambda_1 - b_{ss}|^2} + \cdots + \frac{r}{|\lambda_1 - b_{ss}|^{m'}}$$

As before we obtain correctly ranged components for all components of the B eigenvector $\mathbf{x}$ by using these bounds, and then we can form the A eigenvector $\mathbf{y}$ by means of the equation $\mathbf{y} = P\mathbf{x}$ and convert it to a vector of length 1 in the usual way. As always we need to redo the entire computation at an appropriate higher precision if the required number of correct decimals for the eigenvector components is not obtained.

7.13. LINEAR ALGEBRA USING RATIONAL ARITHMETIC

For the problems listed so far in this chapter, if we require that initial matrices be rational or complex rational instead of real or complex, then the nonsolvable problems become solvable, and solvable problems can be rephrased to eliminate the escapes. We list some revised problems below. In all problems wherever the word *rational* appears, it may be replaced by *complex rational* consistently to obtain another correct version of the problem. When dealing with rationals or complex rationals, we often can use the procedures previously described, but instead of testing whether or not two numbers a, b approximate each other ($a \doteq b$), we test whether or not they equal each other ($a = b$).

SOLVABLE PROBLEM 7.12: Given a set of n linear equations in n unknowns, represented by the matrix equation $A\mathbf{x} = \mathbf{b}$, where A is a rational n-square coefficient matrix, $\mathbf{x}$ is the n-vector of unknowns, and $\mathbf{b}$ the rational n-vector of constants, find as rational numbers the unknowns $x_1, \ldots, x_n$, or else indicate that det A is zero.

The method suggested previously for Solvable Problem 7.2 can be used here, except that it is unnecessary to search a column of A for the element of largest absolute value, and if for any column a nonzero "pivot" or clearing element is not found, then the determinant of A must be zero.

SOLVABLE PROBLEM 7.13: Given the n-square rational matrix A, find the elements of the matrix A^{-1} or else indicate that det $A = 0$.

The method suggested for Solvable Problem 7.3 can be used here with similar changes as for the preceding problem.

Consider the most general rational linear equation problem. We have a set of m linear equations in n unknowns with rational coefficients and constants:

$$
\begin{aligned}
a_{11}x_1 + a_{12}x_2 + \cdots + a_{1n}x_n &= b_1 \\
a_{21}x_1 + a_{22}x_2 + \cdots + a_{2n}x_n &= b_2 \\
\vdots \qquad\quad \vdots \qquad\qquad\quad \vdots \qquad\quad &\;\; \vdots \\
a_{m1}x_1 + a_{m2}x_2 + \cdots + a_{mn}x_n &= b_m
\end{aligned}
$$

Whether there are solutions is to be determined, and if there are, the solution set must be characterized in some way. The equivalent matrix equation is $Ax = \mathbf{b}$ with A the rational coefficient matrix, $\mathbf{x}$ an n-vector of unknowns, and $\mathbf{b}$ a rational m-vector of right side constants.

A rational arithmetic procedure using the various row and column operations we have already introduced could be as follows: We apply the operations exchange row (i, j) and add row multiple (i, j, m) to A, attempting to make zero all the elements of A below the diagonal. Each of these operations is always applied to both A and $\mathbf{b}$, so that the new set of equations defined by the new A and $\mathbf{b}$ arrays is equivalent to the old set. The changing elements of A and components of $\mathbf{b}$ we denote by the symbols a'_{ij} and b'_j, respectively. When working on column j of A, if we find $a'_{jj} = 0$, then we search column j for an element below a'_{jj} that is unequal to zero. If we find $a'_{kj} \neq 0$, then we perform the operation exchange row (j, k) to bring this element into the a'_{jj} position. After we have obtained a nonzero pivot element a'_{jj}, we clear each element a'_{qj} in column j below the pivot with the operation add row multiple $(j, q, -a'_{qj}/a'_{jj})$.

If in column j neither the element a'_{jj} nor the elements below it are unequal to zero, then we search the columns to the right of column j, one by one, attempting to find an element in row j or below that is unequal to zero. If we find $a'_{pq} \neq 0$, we perform the operation exchange col (q, j) to bring this element to the a'_{pj} position, and then we can proceed as before to bring column j to the proper form. The exchange col (q, j) operation of course cannot be performed on b, and to keep the equations represented by our array equivalent to the initial equations, as explained in Section 7.8, we perform the related operation exchange row (q, j) on a certain n-vector initially holding the integers 1 through n in natural order, the vector serving to identify $\mathbf{x}'$, the vector of unknowns. After as many columns of A as possible have been brought to the desired form, the appearance of all the arrays is as follows:

$$
\begin{bmatrix}
a'_{11} & a'_{12} & \cdots & a'_{1r} & \cdots & a'_{1n} \\
 & a'_{22} & \cdots & a'_{2r} & \cdots & a'_{2n} \\
 & & \ddots & \vdots & & \vdots \\
 & & & a'_{rr} & \cdots & a'_{rn} \\
 & & & & 0 & \\
 & & & & & \vdots \\
 & & & & & 0
\end{bmatrix}
\begin{bmatrix}
x'_1 \\ x'_2 \\ \vdots \\ x'_r \\ x'_{r+1} \\ \vdots \\ x'_n
\end{bmatrix}
=
\begin{bmatrix}
b'_1 \\ b'_2 \\ \vdots \\ b'_r \\ b'_{r+1} \\ \vdots \\ b'_n
\end{bmatrix}
$$

If any constant b'_q in the list of constants $b'_{r+1}, \ldots, b'_n$ is unequal to zero, then there are no solutions to the equations, because we have found an equivalent set of equations with one equation being $0 = b'_q$

If all the constants $b'_{r+1}, \ldots, b'_n$ are zero, we can continue on to obtain a representation of the solutions to the equations. The A and $\mathbf{b}$ arrays below row r are no longer needed and are discarded. Working from column r back toward column 1, we gradually form an identity submatrix in the first r columns of A. When working with column j, we apply the operation multiply row $(j, 1/a'_{jj})$ to both A and $\mathbf{b}$ to make a'_{jj} equal 1. Then we subtract appropriate multiples of row j from rows above to clear column j above the unit a'_{jj} element. Our arrays now have the form

$$
[I \quad C'] \begin{bmatrix} \mathbf{x}'_B \\ \mathbf{x}'_{NB} \end{bmatrix} = \mathbf{b}' \tag{7.29}
$$

where I is r-square, the matrix C', a submatrix of A', is of size $r \times (n - r)$, $\mathbf{x}'_B$ and $\mathbf{b}'$ are r-vectors, and $\mathbf{x}'_{NB}$ is an $(n - r)$-vector. The solutions are given by the matrix representation

$$
\mathbf{x}' = \begin{bmatrix} \mathbf{x}'_B \\ \mathbf{x}'_{NB} \end{bmatrix} = \begin{bmatrix} \mathbf{b}' - C'\mathbf{y} \\ \mathbf{y} \end{bmatrix} \tag{7.30}
$$

where the elements of the $(n - r)$-vector $\mathbf{y}$ can be set to any values we please. Here the subscripts B and NB stand for *basic* and *nonbasic*, terms from linear programming. If we set $\mathbf{y}$ to $\mathbf{0}$, we obtain, after undoing the scrambling of $\mathbf{x}'$ components, the solution vector $\mathbf{b}_0$. Similarly, by setting $\mathbf{y}$ to various vectors with all components zero except for a single component equal to 1, we can obtain $n - r$ linearly independent solution vectors, which can be represented as $\mathbf{b}_0 + \mathbf{x}_1, \mathbf{b}_0 + \mathbf{x}_2, \ldots, \mathbf{b}_0 + \mathbf{x}_{n-r}$. It is clear then that any solution to the original set of equations has the form $\mathbf{b}_0$ plus multiples of the vectors $\mathbf{x}_i$.

SOLVABLE PROBLEM 7.14: Given a set of m linear equations in n unknowns, represented by the matrix equation $A\mathbf{x} = \mathbf{b}$, where A is a rational $m \times n$ coefficient matrix, $\mathbf{x}$ is the n-vector of unknowns, and $\mathbf{b}$ a rational m-vector of constants, decide whether or not they have a solution. If they have a solution, determine the rank r of the coefficient matrix, and find a vector $\mathbf{b}_0$ and $n - r$ linearly independent rational vectors $\mathbf{x}_1, \ldots, \mathbf{x}_{n-r}$ such that any real vector $\mathbf{x}$ represents a solution to the equations if and only if it can be written in the form

$$\mathbf{x} = \mathbf{b}_0 + c_1 \mathbf{x}_1 + c_2 \mathbf{x}_2 + \cdots + c_{n-r} \mathbf{x}_{n-r}$$

with real constants c_i.

The set of unknowns x_i', which form x_B', and the set, which form x_{NB}', though apparently fixed by the procedure described, can be varied. If before we start the procedure we scramble the order of the columns of A by doing one or more operations exchange col (i, j) on A, performing the needed related operations on $\mathbf{x}'$, this affects the vectors $\mathbf{x}_B'$ and $\mathbf{x}_{NB}'$ that we obtain.

It is clear that the determination of rank of a rational matrix poses no difficulty.

SOLVABLE PROBLEM 7.15: Detemine the rank of an $m \times n$ rational matrix A.

The solution of the problem of finding eigenvalues and eigenvectors for a rational matrix A requires that we leave the field of rational numbers. Nevertheless, when A is rational, there are no difficulties. We can solve the more difficult problem of finding the Jordan form J of A, in a way that completely solves the eigenvalue–eigenvector problem. We can determine which Jordan blocks of J are associated with any particular eigenvalue λ, and also determine to k decimals the elements of a matrix P, such that $P^{-1} AP = J$. Then a complete set of linearly independent eigenvectors for λ is given by certain columns of P, one column for each Jordan block associated with λ.

SOLVABLE PROBLEM 7.16: Given an n-square rational matrix A, determine to k decimals the elements of the Jordan form J of A, and to k decimals the elements of a complex matrix P, such that $P^{-1} AP = J$.

We can use the Danilevsky procedure of Section 7.11 to find a rational matrix Q, which transforms A to C, where the rational matrix C is composed of a number of companion matrices, $C_1, C_2, \ldots, C_s$, all in diagonal position. The Danilevsky procedure converts to rational computation without difficulty.

From the rational characteristic polynomial of one of the companion submatrices C_i, we form the eigenvalue approximations $\lambda_1^{(i)}, \lambda_2^{(i)}, \ldots, \lambda_q^{(i)}$, with respective multiplicities $m_1^{(i)}, m_2^{(i)}, \ldots, m_q^{(i)}$. From these we construct an appropriate Vandermonde matrix V and then obtain the matrix $P = QV^{-1}$, which transforms A to Jordan form J. If there is more than one companion matrix, the various eigenvalue sets must be compared with each other, this comparison necessarily now being done with range arithmetic. In general, we obtain a list of eigenvalues $\lambda_1, \lambda_2, \ldots, \lambda_t$ with respective multiplicities, but these computed multiplicities need to be verified whenever two eigenvalues from different companion matrices approximate each other ($\doteq$). The characteristic polynomials of all the various companion matrices are multiplied together to obtain the rational characteristic polynomial of the problem matrix, and this polynomial is analyzed using the Euclidean procedure of Chapter 6 to determine the true eigenvalue multiplicity structure. If this differs from the computed multiplicity structure, then the zeros of the various companion matrices must be found to a higher precision and compared anew. Knowing the correct multiplicities for the eigenvalues and knowing the various Jordan blocks that are associated with an eigenvalue λ_i, we obtain a complete set of linearly independent eigenvectors for λ_i by taking as eigenvectors those columns of P with the same column indices as the leading columns of the Jordan blocks associated with λ_i.

7.14. LINEAR PROGRAMMING

Linear programming (or linear optimization) is a mathematical theory for finding the maximum or minimum of a linear expression in certain variables $x_1, x_2, \ldots, x_n$, when the variables are restricted by a set of linear inequalities. Such problems arise in many areas, a prime example being the management of inventory or production. In the next section we will see that the solution of a set of linear interval equations, an important problem that arises frequently in various engineering disciplines, can be posed as a linear programming problem. To prepare for the treatment of that problem we present here the fundamentals of linear programming.

Suppose the linear expression f to be optimized, called the *objective function*, is

$$f = f_1 x_1 + f_2 x_2 + \cdots + f_n x_n + f_0$$

and the linear inequalities that must be satisfied are

$$
\begin{aligned}
a_{11} x_1 + a_{12} x_2 + \cdots + a_{1n} x_n &\leq b_1 \\
a_{21} x_1 + a_{22} x_2 + \cdots + a_{2n} x_n &\leq b_2 \\
\vdots \qquad\qquad\qquad &\quad\ \vdots \\
a_{m1} x_1 + a_{m2} x_2 + \cdots + a_{mn} x_n &\leq b_m
\end{aligned}
\tag{7.31}
$$

The inequalities may be written in matrix form as $Ax \leq \mathbf{b}$. In linear programming problems the variables x_i are restricted to nonnegative values, so the additional inequalities below are always implied

$$x_1 \geq 0, \, x_2 \geq 0, \ldots, x_n \geq 0 \tag{7.32}$$

In analogy with the quadrant concept of the cartesian plane, the points satisfying (7.32) are said to lie in the *first orthant* of the n-space. If in the above inequalities, we prefix some of the variables with a minus sign, we define another orthant of the n-space, there being a total of 2^n orthants.

We assume that all the initial constants a_{ij}, b_i, and f_i of a linear programming problem are rational constants, because the solution method we describe, the *simplex method*, at various steps requires deciding whether certain quantities are positive, negative, or zero, and there is no difficulty making such decisions if we use rational arithmetic.

The first step in the solution procedure is to rewrite each inequality of the set 7.31 as an equation by introducing one extra variable to represent the difference between the two sides of the inequality:

$$
\begin{aligned}
a_{11}x_1 &+ a_{12}x_2 + \cdots + a_{1n}x_n + x_{n+1} = b_1 \\
a_{21}x_1 &+ a_{22}x_2 + \cdots + a_{2n}x_n + x_{n+2} = b_2 \\
&\vdots \qquad\qquad\qquad\qquad\qquad \vdots \\
a_{m1}x_1 &+ a_{m2}x_2 + \cdots + a_{mn}x_n + x_{n+m} = b_m
\end{aligned}
\tag{7.33}
$$

The introduced variables $x_{n+1}, x_{n+2}, \ldots, x_{n+m}$ are called *slack* variables, and like the original variables, they must be nonnegative, so we still have $x_i \geq 0$ for all i. All these relations may be represented in matrix terms as

$$[A \, I]\mathbf{x} = \mathbf{b} \qquad \text{and} \qquad \mathbf{x} \geq \mathbf{0} \tag{7.34}$$

Here the vector $\mathbf{x}$ now has $m + n$ components. The rank of the coefficient matrix $[A \, I]$ is clearly m.

The general solution to the matrix equation of line 7.34 can be found by the method described in the preceding section, which leads to the Equations 7.29 and 7.30. Each possible solution $\mathbf{x}$ determines a point in our $(m+n)$-dimensional space S. Because the vector $\mathbf{x}'_{NB}$ has n components, the set of solution points lies in a certain Euclidean n-dimensional hyperplane H of S. Any point satisfying both relations of line 7.34, called a *feasible point*, lies in the intersection of H with the first orthant of S. The set of feasible points F, if it is not empty or infinite in extent, may be thought of as a kind of generalized polyhedron lying in S. The set F is convex, because if $\mathbf{x}_1$ and $\mathbf{x}_2$ are the vectors defining two feasible points, all points on the line segment joining them are also feasible. The line segment is defined by $\mathbf{x} = (1 - t)\mathbf{x}_1 + t\mathbf{x}_2$, $0 \leq t \leq 1$. We have $[A \, I] \, \mathbf{x}_1 = \mathbf{b}$ and $[A \, I] \, \mathbf{x}_2 = \mathbf{b}$, and $\mathbf{x}_1 \geq \mathbf{0}$, $\mathbf{x}_2 \geq \mathbf{0}$. As t varies, we always have $[A \, I] \mathbf{x} = \mathbf{b}$ and $\mathbf{x} \geq \mathbf{0}$.

The gradient vector of the objective function to be minimized or maximized is always a constant vector. If in F we move in the direction of the gradient, f

increases, while if we move perpendicular to the gradient there is no change in f. If we wish to find the f maximum in F, we can move in any direction as long as we make some progress in the direction of the gradient. The maximum is found at a corner boundary point, that is, a vertex of F, except when there is no maximum and it turns out that we can move as far as we please in the direction of the gradient, F being infinite in extent and allowing unrestricted motion in the direction of the gradient. Accordingly, except for the case in which there is no finite maximum, the corner points of F are the only points we need test to find the maximum of the objective function. It is possible that an entire boundary line segment or higher dimensional face of F consists of maximum points, because the gradient is perpendicular to it, but in this case we also can find a corner boundary point where the maximum is attained. The situation is similar when searching for the minimum of the objective function.

As shown in the preceding section, the m equations in $n + m$ unknowns given by 7.33 have a solution that can be expressed in the matrix form

$$[I \quad C'] \begin{bmatrix} \mathbf{x}'_B \\ \mathbf{x}'_{NB} \end{bmatrix} = \mathbf{b}' \tag{7.35}$$

where the components of the n-vector $\mathbf{x}'_{NB}$ may be set to any value, the components of the m-vector $\mathbf{x}'_B$ being then determined. The explicit equations represented by the matrix equation above are

$$\begin{bmatrix} x'_1 \\ x'_2 \\ \vdots \\ x'_m \end{bmatrix} + \begin{bmatrix} c'_{11} & c'_{12} & \cdots & c'_{in} \\ c'_{21} & c'_{22} & \cdots & c'_{2,n} \\ \vdots & \vdots & \vdots\vdots\vdots & \vdots \\ c'_{m1} & c'_{m2} & \cdots & c'_{m,m} \end{bmatrix} \begin{bmatrix} x'_{m+1} \\ x'_{m+2} \\ \vdots \\ x'_{m+n} \end{bmatrix} = \begin{bmatrix} b'_1 \\ b'_2 \\ \vdots \\ b'_m \end{bmatrix} \tag{7.36}$$

The component variables x'_i of $\mathbf{x}'_B$ are called *basic* variables, the component variables x'_{m+j} of $\mathbf{x}'_{NB}$ being *nonbasic*. Because of the special form of the matrix equation of line 7.34, with the coefficient matrix $[A \ I]$ containing an identity submatrix on the right, the simplest route to the representation above would be to use column exchange operations on the coefficient matrix and related operations on $\mathbf{x}'$ to make the slack variables form the vector $\mathbf{x}'_B$, the matrix C' being then A. We call this the *simple* solution representation. However a solution representation is found, when the vector $\mathbf{x}'_{NB}$ is set to $\mathbf{0}$, the vector $\mathbf{x}'_B$ equals the current $\mathbf{b}'$ vector. We call the point determined this way the *null point* of the solution representation. If it turns out that all the constants b'_i are nonnegative, then the null point is a corner point of the set F of feasible points. This point satisfies all the constraints of 7.34, and is a corner point because if $\mathbf{x}$ is to continue to define a feasible point, the components of $\mathbf{x}'_{NB}$ can change in only one direction, that is, become positive.

The simplex method of solving a linear programming problem starts with any solution representation whose null point is a corner point of F and then locates another solution representation with its null point again a corner point but with the objective function f at the new corner point showing an improved value. This process of f improvement continues until finally a corner point is found at which f is optimum.

If the components of the starting $\mathbf{b}$ vector of 7.34 are all nonnegative, the null point of the simple solution representation is a corner point. When this fortunate situation does not occur, the problem of locating a starting corner point is more difficult, and we deal with it later.

We assume now our linear programming problem requires us to maximize the objective function. (The changes needed in the procedure to locate the minimum of f will be obvious. Or we could find a minimum by using the maximizing procedure on $-f$.) When we are ready to improve the objective function, it must be expressed entirely in terms of the nonbasic variables. Note that if our first corner point was obtained by the simple solution representation, this is automatically the case. To make the process of eliminating basic variables from f easy, it is convenient, before starting the procedure, to write f in the form

$$f + f'_1 x_1 + f'_2 x_2 + \ldots + f'_n x_n = f'_0$$

and to carry the coefficients f'_i as an extra last row of the starting $[A \; I]$ array, except for the coefficient f'_0 which becomes an extra last component of the $\mathbf{b}$ vector. Initially, at the bottom of the $[A \; I]$ array, $f'_j = -f_j$ for $j = 1, \ldots, n$, and $f'_j = 0$ for $j = n + 1, \ldots, n + m$. At the bottom of the $\mathbf{b}$ vector, $f'_0 = f_0$. With this arrangement for the f coefficients, whenever an operation exchange col(i, j) on $[A \; I]$ and $\mathbf{x}'$ is performed, the correct exchange of f'_j coefficients is made also. In the process of bringing the coefficient matrix $[A \; I]$ to the form $[I \; C']$, described in the preceding section, when in a column the elements below a pivot element are cleared to zero, the element in this column belonging to the extra last row gets cleared too. Whenever we arrive at the representation 7.35 or 7.36, the coefficients f'_j appearing below the C' submatrix may be any value, but all coefficients f'_j associated with a basic variable are zero. The value of the objective function at the null point equals the current value of f'_0 in the $\mathbf{b}'$ vector. Similarly, a basic variable x'_k equals b'_k at the null point.

We begin now the description of the simplex cycle at a solution representation given by 7.36. The coefficients of the objective function that correspond to nonbasic variables are tested to find a negative one. These coefficients are $f'_{m+1}, f'_{m+2}, \ldots, f'_{m+n}$. (The column index of any element of the submatrix C' is given as the appropriate column index of the combined $m \times (m + n)$ array.) Suppose f'_{m+q} is negative. The nonbasic variable x'_{m+q} is increased from its zero value to make f larger. As x'_{m+q} increases, the basic variables are affected. The basic variable x_i increases, decreases, or does not change, according as $c'_{i,m+q}$ is negative, positive, or zero, respectively. If $c'_{i,m+q}$ is positive and x'_i decreases, it must not decrease below zero, otherwise the null point of the new solution representation is nonfeasible. The basic variable x'_i becomes zero when x'_{m+q} equals $b'_i/c'_{i,m+q}$. Thus the nonbasic variable x'_{m+q} can be allowed to increase only to the minimum $b'_i/c'_{i,m+q}$ value determined for those positive coefficients c'_{m+q} in column $m + q$. If there happens to be no positive coefficients in this column, then x'_{m+q} can be allowed to increase indefinitely, and the linear programming problem is revealed as a problem without a finite maximum. Suppose a minimum of the $b'_i/c'_{i,m+q}$ values occurs for i equal to r. The nonbasic variable x'_{m+q} is increased to this minimum and becomes basic, and the basic variable x'_r is decreased to zero and becomes nonbasic. We make these changes in our $[I \ C']$, $\mathbf{b}'$, and $\mathbf{x}'$ arrays in the following way: The operation multiply row $(r, \frac{1}{c'_{r,m+q}})$ on $[I \ C']$ and $\mathbf{b}'$ changes the $c'_{r,m+q}$ element to 1, and afterwards a series of appropriate operations, add row multiple (r, j, m) on $[I \ C']$ and $\mathbf{b}'$ clears column $m + q$ of $[I \ C']$ (and also clears f'_{m+q}). Finally exchange of roles of x'_r and x'_{m+q} is accomplished by the operation exchange col(r, $m + q$) on $[I \ C']$ and $\mathbf{x}'$. All these operations are allowable operations on a set of linear equations. Both the starting arrays and the ending arrays of $[I \ C']$, $\mathbf{x}'$, and $\mathbf{b}'$ give the complete set of solutions to the Equations 7.33. The ending arrays are now ready for the next simplex cycle.

When for a solution representation we cannot find any negative coefficients f'_i under the submatrix C', we have located a corner point where f is maximum, and the simplex process terminates. The maximum of f is the current f'_0 value, and the corner point at which this maximum occurs is the null point for this final solution representation. The point is determined when the current nonbasic vector $\mathbf{x}'_{NB}$ equals $\mathbf{0}$, making the current basic vector $\mathbf{x}'_B$ equal the current $\mathbf{b}'$ vector, and we can find the value of the original variables x_i by consulting the current $\mathbf{x}'$ permutation array.

In the simplex cycle, when exchanging the roles of x'_r and x'_{m+q}, if b'_r is positive, the new f'_0 coefficient is larger than the preceding one, because the coefficient f'_{m+q} is negative, so that a positive multiple of b'_r is added to f'_0 when f'_{m+q} is cleared. Because f'_0 is larger than any previous f'_0 value, the corner point determined by the new solution representation is different from any previous corner point. If b'_r is zero, the objective function's f'_0 value stays the same. In this case, it is possible the new corner point is identical to a previous one, and there is a danger that we are in a simplex cycle loop. A number of ways have been proposed to eliminate the possibility of loops, and in our linear programming demonstration program we have used the method proposed by Dantzig, Orden, and Wolfe [3]. With this method we measure "improvement" of the objective function in a more general way, using vector comparison involving additional

f' coefficients instead of just f_0' in the $\mathbf{b}'$ column. At the start of the simplex process, the x_i variables forming the *first* basic vector $\mathbf{x}_B'$ are recorded. Let $x_{\langle 1 \rangle}, \ldots, x_{\langle m \rangle}$ be these variables. The vector $\mathbf{f}'$ to be used in comparisons has $m + 1$ components $f_0', f_{\langle 1 \rangle}', \ldots, f_{\langle m \rangle}'$. Here $f_{\langle i \rangle}'$ denotes the coefficient of f' associated with the specific variable $x_{\langle i \rangle}$. A similar vector $\mathbf{c}_i'$ is defined for any row of the $[I \ C']$ array by taking its components to be $b_i', c_{i,\langle 1 \rangle}', \ldots, c_{i,\langle m \rangle}'$.

Now let $\mathbf{u}$ and $\mathbf{v}$ be two such vectors, and let $\mathbf{0}$ be such a vector with all components zero. We count $\mathbf{u} \succ \mathbf{v}$, if, comparing the components in order and ignoring components that are equal, the first larger component belongs to $\mathbf{u}$. If $\mathbf{u} \succ \mathbf{v}$, we say that the vector $\mathbf{u}$ is "lexicographically" greater than the vector $\mathbf{v}$. Thus $(2, 2, 4, \ldots, 3) \succ (2, 1, 3, \ldots, 5)$. Notice that many rules for inequalities with real numbers have analogous rules for our vector relation: From $\mathbf{u} \succ \mathbf{v}$, it follows that $\mathbf{u} - \mathbf{v} \succ \mathbf{0}$. It also follows that $a\mathbf{u} \succ a\mathbf{v}$, where a is any positive number, and that $\mathbf{u} + \mathbf{w} \succ \mathbf{v} + \mathbf{w}$ where $\mathbf{w}$ is any vector. Also, from $\mathbf{u}_1 \succ \mathbf{v}_1$ and $\mathbf{u}_2 \succ \mathbf{v}_2$, it follows that $\mathbf{u}_1 + \mathbf{u}_2 \succ \mathbf{v}_1 + \mathbf{v}_2$.

In a simplex cycle, previously when f_{m+q}' was negative, we chose a row r such that $b_r'/c_{r,m+q}'$ was the minimum of the quantities $b_i'/c_{i,m+q}'$ with i such that $c_{i,m+q}'$ is positive. However, there could be a tie for mimimum, and we revise our selection system to make the choice unique. Now we choose row r if the vector $\frac{1}{c_{r,m+q}'}\mathbf{c}_r'$ is the lexicographic minimum of the vectors $\frac{1}{c_{i,m+q}'}\mathbf{c}_i'$ with i such that $c_{i,m+q}'$ is positive.

Note that at the beginning of the first simplex cycle, the last m components of the vectors $\mathbf{c}_i'$ are taken from the matrix I. This implies that a comparison tie among the vectors $\frac{1}{c_{i,m+q}'}\mathbf{c}_i'$ is impossible and that $\mathbf{c}_i' \succ \mathbf{0}$ for all i. We show that if these two conditions exist at the start of a simplex cycle, they exist at the end of the cycle, and so always hold. First note that a comparison tie implies that the m vectors $\mathbf{c}_i'$ are linearly dependent. But if these vectors are linearly independent initially, this is not changed by any row or column operations performed in a simplex cycle, so a comparison tie at the conclusion of a simplex cycle is again not possible. Next note that when column $m + q$ is cleared by adding multiples of row r to other rows, if $c_{i,m+q}'$ is negative, a positive multiple of row r is added to row i. Because $\mathbf{c}_r' \succ \mathbf{0}$, for the new vector $\mathbf{c}_i'$ we have $\mathbf{c}_i' \succ \mathbf{0}$ if this relation was true for the old vector $\mathbf{c}_i'$. If $c_{i,m+q}'$ is positive, the multiple $\frac{c_{i,m+q}'}{c_{r,m+q}'}$ of row r is subtracted from row i, but because we had $\frac{1}{c_{i,m+q}'}\mathbf{c}_i' \succ \frac{1}{c_{r,m+q}'}\mathbf{c}_r'$, implying

$\mathbf{c}_i' \succ \frac{c_{i,m+q}'}{c_{r,m+q}'}\mathbf{c}_r'$, we obtain $\mathbf{c}_i' \succ \mathbf{0}$ for the new vector $\mathbf{c}_i'$. The multiplication of row r by the positive constant $\frac{1}{c_{r,m+q}'}$ maintains the relation $\mathbf{c}_r' \succ \mathbf{0}$. Thus $\mathbf{c}_r' \succ \mathbf{0}$ for all i at the conclusion of the simplex cycle.

It is clear now that after a positive multiple of $\mathbf{c}'_r$ is added to $\mathbf{f}'$, the new $\mathbf{f}'$ vector is lexicographically greater than the old one. This implies that the new solution representation is different from any preceding solution representation, and simplex cycle loops now are impossible.

Now we address the problem of locating an initial corner point when the simple method of obtaining a solution representation does not determine one. Assume that we have obtained *any* solution representation, possibly the simple one, and that some of the components b'_i are negative, so our representation's null point is not feasible. We call a basic variable x'_i negative or nonnegative according as its current b'_i component is negative or nonnegative. We collect in a list L the index i of any negative basic variable x'_i. We say a basic variable is on the list L if its index is in L. The general idea is to adapt the simplex procedure to the problem of finding the largest value of the linear function equal to the sum of basic variables on the list L. The list L varies as we proceed, with a basic variable x'_i being discarded from L as soon as it becomes nonnegative. When the list is empty, our solution representation determines a feasible null point. The varying linear function f, defined as the sum of all basic variables currently on the list L, has corresponding to a nonbasic variable x'_{m+j} a coefficient f'_{m+j} that equals $\Sigma_{i \in L} \, c'_{i,m+j}$. Because we want to increase f, we test these coefficients to find a negative one, say f'_{m+q}. As before, we wish to make the nonbasic variable x'_{m+q} positive to increase our temporary objective function. Here the increase is limited to the minimum of the quantities $b'_i/c'_{i,m+q}$ for i not on L with $c'_{i,m+q}$ positive, because we do not want any nonnegative basic variables to become negative. However, there may be no indices i satisfying the conditions given. In this case we can allow x'_{m+q} to increase to the maximum of the *positive* quantities $b'_i/c'_{i,m+q}$ for i on L with $c'_{i,m+q}$ negative, because then all the corresponding negative basic variables x'_i in this group become nonnegative, and nonnegative basic variables stay nonnegative. Once the bound on x'_{m+q} is determined, the process of altering the $[I \ C']$ and $\mathbf{b}'$ arrays and exchanging the variable x'_{m+q} with a basic variable is the same as before. After we have a new solution representation, the entire process is repeated, and continues until the list L is empty and a feasible null point has been determined. If it ever happens that none of the f'_{m+j} coefficients are negative, the linear programming problem is revealed as one that has no feasible points.

7.15. INTERVAL LINEAR EQUATIONS AGAIN

We reconsider the problem of solving the interval equations 7.2 when all interval endpoints are rational numbers. The problem then is:

SOLVABLE PROBLEM 7.17: Given a system of n linear interval equations in n unknowns, represented by the matrix equation $A\mathbf{x} = \mathbf{b}$ with A and $\mathbf{b}$ having interval elements with rational endpoints, either give optimum rational upper and lower bounds for the unknowns x_i, or indicate that the matrix A is singular.

Often instead of rational bounds, what we want are optimum upper and lower x_i bounds to k correct decimals. These of course are easily obtained from the rational bounds.

The interval elements of A and $\mathbf{b}$ can be expressed in either the endpoint notation or the mid-wid notation. If for an interval elements a_{ij} we use the notation $a_{ij}^{(m)}$ for the element's mid and $a_{ij}^{(w)}$ for its wid, and similar notation for the interval element b_i, then for all i, j

$$a_{ij} = [a_{ij}^{(m)} - a_{ij}^{(w)}, a_{ij}^{(m)} + a_{ij}^{(w)}] \qquad b_i = [b_i^{(m)} - b_i^{(w)}, b_i^{(m)} + b_i^{(w)}]$$

We use $A^{(m)}$ for the matrix of mids and $A^{(w)}$ for the matrix of wids and similarly $\mathbf{b}^{(m)}$ and $\mathbf{b}^{(w)}$ for the mid and wid vectors. The following result is from Oettli and Prager [13].

THEOREM 7.6: *A necessary and sufficient condition for a real n-vector* $\mathbf{x}$ *to specify a solution point to the interval matrix equation* $A\mathbf{x} = \mathbf{b}$ *is*

$$\left| A^{(m)}\mathbf{x} - \mathbf{b}^{(m)} \right| \le A^{(w)}|\mathbf{x}| + \mathbf{b}^{(w)} \tag{7.37}$$

or, equivalently,

$$-A^{(w)}|\mathbf{x}| - \mathbf{b}^{(w)} \le A^{(m)}\mathbf{x} - \mathbf{b}^{(m)} \le A^{(w)}|\mathbf{x}| + \mathbf{b}^{(m)} \tag{7.38}$$

The necessity of the condition is easy to show. If the n-vector $\mathbf{x}$ specifies a solution point, there is an n-square matrix $\hat{A}$ and an n-vector $\hat{\mathbf{b}}$ such that $\hat{A}\mathbf{x} = \hat{\mathbf{b}}$, and all elements $\hat{a}_{ij}$ and $\hat{b}_i$ are contained in respective intervals a_{ij} and b_i, that is,

$$A^{(m)} - A^{(w)} \le \hat{A} \le A^{(m)} + A^{(w)} \qquad \text{and} \qquad \mathbf{b}^{(m)} - \mathbf{b}^{(w)} \le \hat{\mathbf{b}} \le \mathbf{b}^{(m)} + \mathbf{b}^{(w)}$$

We have

$$\begin{aligned}
\left| A^{(m)}\mathbf{x} - \mathbf{b}^{(m)} \right| &= \left| A^{(m)}\mathbf{x} - \mathbf{b}^{(m)} - (\hat{A}\mathbf{x} - \hat{\mathbf{b}}) \right| \\
&= \left| (A^{(m)} - \hat{A})\mathbf{x} - (\mathbf{b}^{(m)} - \hat{\mathbf{b}}) \right| \\
&\le \left| A^{(m)} - \hat{A} \right| |\mathbf{x}| + \left| \mathbf{b}^{(m)} - \hat{\mathbf{b}} \right| \\
&\le A^{(w)}|\mathbf{x}| + \mathbf{b}^{(w)}
\end{aligned}$$

To show the sufficiency of the condition, suppose the matrix relation 7.37 is true for a certain n-vector $\mathbf{x}$. We must find an acceptable matrix $\hat{A}$ and an acceptable vector $\hat{\mathbf{b}}$ such that $\hat{A}\mathbf{x} = \hat{\mathbf{b}}$. Set the matrix $\hat{A}$ equal to $A^{(m)} + A^{(\delta)}$, and the vector $\hat{\mathbf{b}}$ equal to $\mathbf{b}^{(m)} + \mathbf{b}^{(\delta)}$. We need to determine $A^{(\delta)}$ and $\mathbf{b}^{(\delta)}$ so that

$$|A^{(\delta)}| \leq A^{(w)} \qquad \text{and} \qquad |\mathbf{b}^{(\delta)}| \leq \mathbf{b}^{(w)} \tag{7.39}$$

From the matrix equation $\hat{A}\mathbf{x} = \hat{\mathbf{b}}$ we obtain

$$A^{(\delta)}\mathbf{x} - \mathbf{b}^{(\delta)} = \mathbf{y} = -(A^{(m)}\mathbf{x} - \mathbf{b}^{(m)})$$

where the components of the vector $\mathbf{y}$ are determined by the right-hand side of this equation. The equation for the i-th component of $\mathbf{y}$ is

$$\sum_{j=1}^{n} a_{ij}^{(\delta)}x_j - b_i^{(\delta)} = y_i \tag{7.40}$$

The elements of $A^{(\delta)}$ and $\mathbf{b}^{(\delta)}$, which appear in this equation, do not appear in the equation for any other component of $\mathbf{y}$, so we are free to set these components as we please to satisfy our equation. If y_i is zero, we take $b_i^{(\delta)} = 0$ and $a_{ij}^{(\delta)} = 0$ for all j, and the relations 7.39 are satisfied. If y_i is nonzero we take $a_{ij}^{(\delta)} = t \cdot a_{ij}^{(w)} \operatorname{sgn} x_j$ and $b_i^{(\delta)} = -t \cdot b_i^{(w)}$, where t is a nonzero parameter. Equation 7.40 now has the form

$$t\left(\sum_{j=1}^{n} a_{ij}^{(w)}|x_j| + b_i^{(w)}\right) = y_i$$

If the parameter t is set to 1, the left side of the equation above is

$$\sum_{j-1}^{n} a_{ij}^{(w)}|x_j| + b_i^{(w)} = \rho_i$$

The inequality 7.38 implies $-\rho_i \leq -y_i \leq \rho_i$, so the positive quantity ρ_i is not less than $|y_i|$. Thus we can take t equal to y_i/ρ_i to satisfy 7.40 with the relations 7.39 true. Thus in all cases, we can determine an appropriate matrix $\hat{A}$ and vector $\hat{\mathbf{b}}$.

The vectors $\mathbf{x}$ that can satisfy Equation 7.37 frequently lie in just one orthant, but it is possible that these vectors stretch over several orthants. For instance, consider the equations below, an example given by Hansen [7]. The solutions to these equations lie in three of the four possible orthants, or quadrants in this case, because the vector space is of dimension 2.

$$[2, 3]x_1 + [0, 1]x_2 = [0, 120]$$

$$[1, 2]x_1 + [2, 3]x_2 = [60, 240] \tag{7.41}$$

When there are solution points in more than one orthant, in each orthant the set of solutions is convex, as will be clear shortly, and these convex parts join together to form a set which in general is not convex.

To find the relations that apply in a particular orthant Q, let D_Q be a diagonal matrix with its i-th diagonal element chosen to be $+1$ or -1 according as x_i is positive or negative, respectively, in the orthant. Then in this orthant Q we have $D_Q \mathbf{x} = |\mathbf{x}|$, and $\mathbf{x} = D_Q|\mathbf{x}|$. The inequality 7.38 now may be written as

$$-A^{(w)}|\mathbf{x}| - \mathbf{b}^{(w)} \leq A^{(m)}D_Q|\mathbf{x}| - \mathbf{b}^{(m)} \leq A^{(w)}|\mathbf{x}| + \mathbf{b}^{(w)}$$

or, equivalently, as

$$(A^{(m)}D_Q - A^{(w)})|\mathbf{x}| \leq \mathbf{b}^{(m)} + \mathbf{b}^{(w)}$$
$$(-A^{(m)}D_Q - A^{(w)})|\mathbf{x}| \leq -\mathbf{b}^{(m)} + \mathbf{b}^{(w)}$$
(7.42)

The set of solution vectors $\mathbf{x}$ in this orthant is convex because if $\mathbf{x}_1$ and $\mathbf{x}_2$ are two solution vectors in the orthant satisfying the inequalities above, then so also does $\mathbf{y} = (1 - t)\mathbf{x}_1 + t\mathbf{x}_2$ for any choice of the parameter t in the interval [0, 1]. The equation for $\mathbf{y}$ defines a line segment in the orthant joining the point defined by $\mathbf{x}_1$ to the point defined by $\mathbf{x}_2$.

To find the largest value of a component $|x_i|$ of $|\mathbf{x}|$ for all solution vectors $\mathbf{x}$ in the orthant Q, we can solve the linear programming problem with the $2n$ linear constraints represented by the two vector inequalities above, and the nonnegativity constraints

$$|x_k| \geq 0, \qquad k = 1, 2, \ldots, n$$

for the maximum of the objective function f equal to $|x_i|$. The smallest value for $|x_i|$ is obtained by solving the identical linear programming problem except that we find the minimum of the objective function.

To simplify our notation we now replace $|x_i|$ everywhere by $\hat{x}_i$. After we assign a slack variable $\hat{x}_{n+k}$ to the k-th component inequality obtained from 7.42 and convert it to an equation, we increase the number of variables to $3n$, and the constraint equations have the matrix representation

$$\begin{bmatrix} A^{(m)}D_Q - A^{(w)} & I & 0 \\ -A^{(m)}D_Q - A^{(m)} & 0 & I \end{bmatrix} \hat{\mathbf{x}} = \begin{bmatrix} \mathbf{b}^{(m)} + \mathbf{b}^{(w)} \\ -\mathbf{b}^{(m)} + \mathbf{b}^{(w)} \end{bmatrix}$$
(7.43)

where $\hat{\mathbf{x}}$ is a $3n$-vector with components $\hat{x}_i$. To start the simplex cycle we need a basic feasible point, that is, a division of the $3n$ variables into $2n$ basic and n nonbasic defining a corner point satisfying all constraints, in the usual form

$$\begin{array}{l}
\hat{x}'_1 + c'_{1,2n+1}\hat{x}'_{2n+1} + \cdots + c'_{1,3n}\hat{x}'_{3n} = b'_1 \\
\hat{x}'_2 + c'_{2,2n+1}\hat{x}'_{2n+1} + \cdots + c'_{2,3n}\hat{x}'_{3n} = b'_2 \\
\vdots \qquad\qquad\qquad\qquad\qquad \vdots \quad\quad \vdots \\
\hat{x}'_{2n} + c'_{2n,2n+1}\hat{x}'_{2n+1} + \cdots + c'_{2n,3n}\hat{x}'_{3n} = b'_{2n}
\end{array}$$
(7.44)

Here $\hat{x}'_1, \hat{x}'_2, \ldots, \hat{x}'_{2n}$ are basic variables and the remainder are nonbasic.

The algorithm to bound the components of **x** can now be described. For each component of x_i of **x**, a pair of variables, "current upper bound" and "current lower bound," is maintained and repeatedly improved as computation proceeds investigating various orthants of our n-space. A starting orthant Q is found by solving the equation $A^{(m)}\mathbf{x} = \mathbf{b}^{(m)}$ for **x**, and using the **x** vector to determine Q and the matrix D_Q. If a component x_j of the solution vector **x** is zero, the corresponding j-th diagonal element of D_Q is taken arbitrarily as $+1$. This fixes the starting orthant's Equations 7.43, and a starting basic feasible point can be found by taking the simple solution representation, followed by the process described in the previous section for converting a solution representation to a feasible solution representation.

Each orthant that is investigated requires a separate determination of the orthant's largest and smallest value for each component $\hat{x}_j = |x_j|$. A task queue L is needed listing the orthants to be investigated. Each task queue entry, besides giving the various x_i signs defining the orthant, specifies the $2n$ basic variables to be used to form a starting feasible solution representation and for which components $\hat{x}_i$ the minimum search should be skipped, the minimum being known to be zero. Initially, when computation begins with the starting orthant the task queue is empty. Let us designate the orthant currently being investigated as Q_C. Initially Q_C is the starting orthant.

As mentioned, for the starting orthant, an initial feasible solution representation is found by computation. For any other orthant investigated, a list of the basic variables that generate a starting feasible solution appears on the task queue entry, and this information is used to generate the solution representation. For such orthants the method of generating the starting $[I\ C']$, $\hat{\mathbf{x}}'$, and $\mathbf{b}'$ arrays is as follows. By column exchanges, the first $2n$ columns of the coefficient matrix are made to correspond to the designated basic variables. These columns are linearly independent. Starting with the first column and proceeding in sequence to column $2n$, in each column a nonzero element is found, brought to the diagonal position by an operation exchange row(i,j), converted to a 1 by an operation multiply row(i,m), and used to clear the column by operations add row multiple(i,j,m). The appropriate operations on $\hat{\mathbf{x}}'$ and $\mathbf{b}'$ always are performed also, and on completion of the process a feasible solution representation is obtained.

The objective functions $f = \hat{x}_i$, $i = 1, 2, \ldots, n$ are maximized one after the other, and then these same objective functions are minimized one after the other. Each objective function problem begins operations with the final solution representation and associated corner point obtained by the preceding problem, except, of course, for the maximization of the first objective function, which begins with the specified solution representation, or in the case of the starting orthant, the computed solution representation.

We consider the maximization problems first. A separate computation of an objective function $f = \hat{x}_i$ does not have to be made whenever the variable $\hat{x}_i$ is a basic variable. For suppose $\hat{x}_i$ is the basic variable $\hat{x}'_p$ defined by row p of 7.44. Then, as with an objective function, we look in row p for a coefficient $c'_{p,2n+k}$ that is negative. If none is found, we are done and the maximum equals b'_p. If a negative coefficient $c'_{p,2n+q}$ is found, we search column $2n + q$ to find row indices s for which $c'_{s,2n+q}$ is positive, computing then the quantity $b'_s / c'_{s,2n+q}$. Let r be the row index for which this quantity is least (or tied for least). We manipulate the elements of $[I\ C']$ by allowable operations to make the basic variable $\hat{x}'_r$ nonbasic and the nonbasic variable $\hat{x}'_{2n+q}$ basic. Here the unboundedness of the objective function $\hat{x}_i$ is detected if no positive elements $c'_{s,2n+q}$ are found in column $2n + q$. This indicates that the interval coefficient matrix A is singular and computation ceases. If $\hat{x}_i$ should be found to be a nonbasic variable, say $\hat{x}'_{2n+t}$, then first we proceed as if column $2n + t$ had been determined as the column to change for an objective function, making our variable basic by the process just described, and then continuing as before.

The procedure for finding the minimum of a variable $\hat{x}_i$ is similar if $\hat{x}_i$ is a basic variable $\hat{x}'_p$, except that we look for positive coefficients $c'_{p,2n+k}$ instead of negative ones. If $\hat{x}_i$ is nonbasic, we are done, because the minimum is identified as zero. Whenever a zero value is obtained for a variable $\hat{x}_i$, this signals that the orthant Q_N, identical to Q_C except for the opposite x_i sign, must be investigated. The task queue list L is searched to see if Q_N is already on the list. If it is, the minimum problem for $\hat{x}_i$ in the Q_N entry is set to be skipped. If it is not on the list, then a Q_N entry is added to the end of the task queue with the $\hat{x}_i$ minimum problem set to be skipped. In this case the starting basic variables for Q_N are assigned the Q_C basic variables, which give the zero value for $\hat{x}_i$. (We may assume $\hat{x}_i$ is nonbasic because it is made nonbasic should it happen not to be so.) Note that D_{Q_C} and D_{Q_N} differ only in the i-th diagonal element, so the array 7.43 for Q_N and the array for Q_C differ only in column i. Suppose the set of equations 7.43 for Q_C were supplied and the current basic variables were designated. By going through the procedure described earlier for obtaining a starting feasible solution, we would obtain the current solution representation without column i being invoked in the process of generating the identity submatrix I of $[I\ C']$, and our null point P is feasible. If these same basic variables are specified for the equations corresponding to Q_N, a solution representation is obtained that has the same null point P, and so the solution is feasible.

After the maxima and minima of all variables $\hat{x}_i$ have been determined for the current orthant Q_C, these are converted to x_i values by attaching the Q_C orthant signs and used to update the current upper and lower bounds for the components of $\mathbf{x}$. After the last orthant has been investigated, the final values of the current upper and current lower x_i bounds give the desired $\mathbf{x}$ interval bounds.

We now consider solving the interval equations 7.42 without the requirement that all interval endpoints be rational.

SOLVABLE PROBLEM 7.18: Given a system of n linear interval equations in n unknowns, represented by the matrix equation $A\mathbf{x} = \mathbf{b}$ with A and $\mathbf{b}$ having interval elements with real endpoints, give to k decimals optimum upper and lower bounds for the unknowns x_i, indicate that the interval matrix A is singular, or indicate that $|\det A| < 10^{-k}$ when all A elements are set to the right endpoints of their intervals.

A solution procedure here could be as follows: We obtain k decimal approximations to the endpoints of all intervals a_{ij} and b_i. A k-decimal approximation to a quantity α defines an interval with rational endpoints within which the quantity α must lie. Thus from the k decimal interval endpoint values for a_{ij}, we can define an interval with rational endpoints containing a_{ij} and another interval with rational endpoints contained within the interval a_{ij}, and similarly for b_i. Thus if $k = 3$ and we obtain for a_{ij} the value $[2.333^\sim, 2.666^\sim]$, then a_{ij} is contained in $[2.3335, 2.6665]$ and contains $[2.3325, 2.6655]$.

Using the rational solution procedure described, we can solve the interval problem using the larger rational intervals and solve it again using the smaller rational intervals. If the coefficient matrix is found to be singular using the smaller intervals, then the coefficient matrix is singular for the original problem. If we obtain solutions for both problems, then we can determine if the two sets of rational bounds obtained for x_i allow us to give k decimal bounds to the original problem. If not, we increase by a few places the number of decimals used to obtain the approximations to the interval endpoints and try again. The only time the escape is considered is when we obtain a solution using the smaller intervals but find a singular coefficient matrix using the larger interval. In this case we switch to the problem of determining the determinant specified in Problem 7.18. (Right-hand endpoints are specified for definiteness; left-hand endpoints could have been specified instead.) If the outcome is not satisfactory, we proceed as with the failure of the regular procedure. Eventually we must succeed in satisfying the problem's requirements.

EXERCISES

1. Call equat and solve the equations

$$2x_1 + x_2 = 17$$

$$x_1 + 2x_2 = 19$$

See the form of the answers supplied when this problem is supplied to the rational equation solving program by calling r_equat < equat.log. Try also c_equat < equat.log.

2. Call i_equat1 and solve to four decimals the interval equations 7.3. Check that the optimum bounds given in the text are obtained.

3. Call i_equat1 and attempt to solve to five decimals the interval equations 7.41, which appeared as an example in a paper by Hansen [7]. The interval coefficients and constants for this problem are too wide for i_equat1. Feed the problem to i_equat by calling i_equat < i_equat1.log to obtain the interval solution.

4. Call eigen and find to 10 decimals the eigenvalues and eigenvectors of the matrix below.

$$\begin{bmatrix} 2 & 4 & 4 \\ 0 & 3 & 1 \\ 0 & 1 & 3 \end{bmatrix}$$

For one of the eigenvalues two near eigenvectors are obtained instead of eigenvectors.

5. Call r_eigen < eigen.log to feed the problem of the preceding exercise to the rational matrix program, and note that this time eigenvectors are obtained.

6. Call linpro and solve the linear programming problem of maximizing the function $4x_1 + x_2$ subject to the constraints

$$2x_1 + 7x_2 \leq 8$$
$$x_1 + 2x_2 \leq 3$$
$$x_i \geq 0, i = 1, 2$$

7. Solve the preceding linear programming problem with the first inequality changed to an equation. The program linpro handles a constraint equation $2x_1 + 3x_2 = 4$ by converting it into two constraint inequalities $2x_1 + 3x_2 \leq 4$ and $2x_2 + 3x_2 \geq 4$.

NOTES

N7.1. Two general references on linear algebra are the books by Gantmacher [5] and by Hohn [10]. Two references emphasizing computation methods of linear algebra are the books by Golub and Van Loan [6] and by Faddeev and Fadeeva [4].

N7.2. Two texts on linear programming that have appeared in the last decade are the books by Brickman [2] and by Ignizio and Cavalier [11].

N7.3. The method for solving linear interval equations using partial derivative bounds given in Section 7.5 is based on a method described by Hansen [7].

N7.4. The linear programming method for solving linear interval equations with rational endpoints [1] that is described in Section 7.15 is a generalization of a method proposed by Oettli [13] for the case in which the **x** solution points lie in a single orthant. Other methods for handling this single orthant case are given in Neumaier's book [12]. A general method for treating linear interval equations, with no restrictions placed on the **x** solution, was found by Rohn [15] and is described in Neumaier's book.

REFERENCES

1. Aberth, O., The solution of linear interval equations by a linear programming method, *Linear Algebra Appl.* **259** (1997), 271–279.
2. Brickman, L., *Mathematical Introduction to Linear Programming and Game Theory*, Springer Verlag (Series: Undergraduate Texts in Mathematics), New York, 1989.
3. Dantzig, G. B., Orden, A., and Wolfe, P., The generalized simplex method for minimizing a linear form under linear inequality restraints, *Pacific J. Mathematics* **5** (1955), 183–195.
4. Faddccv, D. K. and Faddccva, V. V., *Computational Methods of Linear Algebra*, W. H. Freeman, San Francisco, 1963.
5. Gantmacher, F. R., *Theory of Matrices, Vols. 1 and 2*, Chelsea Publishing Co., New York, 1960.
6. Golub, G. H. and Van Loan, C. E., *Matrix Computations, 2nd Edit.*, Johns Hopkins University Press, Baltimore, 1989.
7. Hansen, E., On the solution of linear algebraic equations with interval coefficients, *Linear Algebra Appl.* **2** (1969), 153–165.
8. Hansen, E., *A generalized interval arithmetic*, in *Interval Mathematics, K. Nickel editor*, Springer Verlag, New York, 1975, pp 7–18.
9. Hansen, E., Computing zeros of functions using generalized interval arithmetic, *Interval Computations* **3** (1993), 3–28.
10. Hohn, F., *Elementary Matrix Algebra, 3rd Edit.*, Macmillan, New York, 1973.
11. Ignizio, J. P. and Cavalier, T. M., *Linear Programming*, Prentice Hall (International Series in Industrial and Systems Engineering), Englewood Cliffs, NJ, 1994.
12. Neumaier, A., *Interval Methods of Systems of Equations, Encyclopedia of Mathematics and its Applications*, Cambridge University Press, Cambridge, 1990.
13. Oettli, W, On the solution set of a linear system with inaccurate coefficients, *SIAM J. Numer. Anal.* **2** (1965), 115–118.

14. Oettli, W. and Prager, W., Compatability of approximate solution of linear equations with given error bounds for coefficients and right-hand sides, *SIAM J. Numer. Anal.* **2** (1965), 291–299.

15. Rohn, J., Systems of linear interval equations, *Linear Algebra Appl.* **126** (1989), 39–78.

VIII

OPTIMIZATION PROBLEMS

8.1. FINDING A FUNCTION'S EXTREME VALUES

A common problem is that of finding the maximum or minimum of a function $f(x_1, x_2, \ldots, x_n)$ in a specified region of n-space. In Section 5.9, elementary regions were found to be convenient search regions for zeros, and we will take our optimization problem to be that of locating the minimum of an elementary function f in a specified elementary region R. The problem of finding the maximum of f is equivalent to the problem of finding the minimum of $-f$.

If the minimum of f occurs at an interior point of R, then the f gradient is $\mathbf{0}$ if it is defined at the point. If the minimum of f occurs at a boundary point, the f gradient can be any value. Usually one is interested not only in the value of the minimum but also in the argument or arguments where the minimum is attained. However, locating to k decimals an argument point where the minimum occurs is a nonsolvable problem, as is seen by considering the case of the function $f_c(x) = cx$ with the region R taken as the interval $[-1, 1]$. If the parameter c is zero, the minimum of $f_c(x)$ occurs at any argument in $[-1, 1]$; if c is positive, the minimum occurs at $x = -1$; and if c is negative, the minimum occurs at $x = 1$. A program that could find to k decimals a point at which $f_c(x)$ is minimum would let us infer that $c \geq 0$ if the k decimal value supplied were negative or zero and would let us infer that $c \leq 0$ if the value supplied were positive. However, this contradicts Nonsolvable Problem 3.8.

Thus we must retreat from attempting to locate an argument where the maximum occurs. A reasonable alternate problem is

SOLVABLE PROBLEM 8.1: For an elementary function $f(x_1, x_2, \ldots, x_n)$ defined in an elementary region R of n-space, determine the k decimal minimum value of f in R, and give to k decimals all f arguments in R where f has a matching k decimal value, or else give to k decimals an **x** box within which all such arguments must lie.

If several f arguments are supplied, no claim is made that the minimum occurs at any of these arguments. For k larger, a more accurate minimum value is obtained and, possibly, a shorter list of arguments. The box escape is for difficult cases like the function f_c with c close to zero.

We give the procedure first for the case where the elementary region is a box B defined by n intervals with exact endpoints. Later the procedure is revised to handle any elementary region. If the intervals defining a box have exact endpoints, the box can be divided into two subboxes of the same type by bisecting one of these intervals. After a long series of such divisions, we still are able to decide without error whether any two subboxes are touching.

The first step in finding the minimum is to divide B into a list of subboxes such that f can be evaluated without a series error over each of the subboxes. This is a task queue computation with the queue initially holding just the box B. The queue cycle consists of attempting to evaluate f to degree 0 with formal interval arithmetic over the leading queue subbox. If there is a series error, the leading box is divided into two by bisecting its largest dimension interval, and the two subboxes replace it on the queue. If an f interval is obtained, the leading box is added to a list L, initially empty. Eventually the task queue is empty, and the list L has our required B disection.

Combining ideas from computation procedures suggested by Moore [4, 5], Skelboe [9], and Ichida-Fujii [3], we obtain the k-decimal bound on the minimum of f efficiently as follows. For each subbox $B^{(S)}$ on L, if $B^{(S)}$ has no boundary points of B, a degree 1 series evaluation of f is attempted. If this is obtained, then we have intervals for all the partial derivatives of f over $B^{(S)}$. If any of these partial derivative intervals do not contain 0, then it is certain that the minimum of f does not occur in $B^{(S)}$, and the subbox is discarded. When a degree 1 evaluation is not obtained, a degree 0 computation of f is, which is all that is attempted for any box $B^{(S)}$ containing boundary points of B. Thus for these subboxes, and for all the nondiscarded subboxes with degree 1 evaluations, we obtain an interval $m_S \pm w_S$ for f over $B^{(S)}$. The f minimum over $B^{(S)}$ may be as large as m_S, the f value at the centerpoint of $B^{(S)}$, and may be as small as $m_S - w_S$. Thus the f minimum over $B^{(S)}$ is somewhere in the interval $[m_S - w_S, m_S]$.

The surviving subboxes on the list L are now arranged in order as follows: subbox B_S precedes subbox $B_{S'}$ on the list if $m_S - w_S \lessdot m_{S'} - w_{S'}$, or, in the case where $m_S - w_S \doteq m_S' - w_S'$, if $w_S \gg w_S'$. An initial interval bound $[M_1, M_2]$ for the minimum of f may be taken as $[m_{S_0} - w_{S_0}, m_{S_0}]$ where B_{S_0} is the first subbox on the list L.

The following process is repeatedly performed to reduce the width of the interval $[M_1, M_2]$. The first subbox on L is removed and divided into two subboxes by bisecting its largest dimension interval. Each of these two subboxes goes through the f evaluation procedure described and, if not discarded, is placed in its appropriate position on the L list. The new value of M_1 is $m_{S_0} - w_{S_0}$ where B_{S_0} is the new leading element on L. Only if $m_{S_0} \ll M_2$ is M_2 reduced to the new value $M_2 = m_{S_0}$, and in this case a pass is made through the list L to discard any subbox B_S for which $m_S - w_S \gg M_2$.

The minimum of f equals the ranged number $\frac{1}{2}(M_1 + M_2) \oplus \frac{1}{2}(M_2 - M_1)$. The process of bisecting the first element of L continues until an f minimum accurate to k decimals is obtained.

The next step is to use the surviving members of the final L list to find where in B the minimum occurs. The L subboxes are arranged into connected sets of subboxes, and for each set a containing box $B^{(C)}$ is constructed, just large enough to hold the members of the set. If each box $B^{(C)}$ is small enough to fit within the tilde-box of its centerpoint to k decimals, then we satisfy the requirements of Problem 8.1 by displaying all the centerpoints to k decimals.

If some of the $B^{(C)}$ boxes fail this criterion, the list L is processed anew obtaining f's minimum value to more decimals in an attempt to get container boxes sufficiently small that centerpoints can be displayed. If there is no progress toward this goal, the escape exit is used. That is, we construct a containing box just large enough to hold all the subboxes of the final list L and display to k decimals the endpoints of its defining intervals.

For a general elementary region R, we use the $\mathbf{u}$ unit box $B^{(\mathbf{U})}$, and map the points of $B^{(\mathbf{u})}$ into R, using the functions defining the elementary region, as described in Section 5.9, and in this way obtain f as a function of $\mathbf{u}$. Here for any subbox $B^{(S)}$ of $B^{(\mathbf{u})}$, the evaluation of f over $B^{(S)}$ requires two steps. First we obtain intervals for all x_i variables, these intervals defining an $\mathbf{x}$ box. Then with all x_i primitives set accordingly, we obtain either the degree 1 or the degree 0 series expansion for f. After the f minimum is found and the final L list is processed to locate arguments, the $B^{(\mathbf{u})}$ subboxes determine the container boxes $B^{(C)}$, but it is the size of their $\mathbf{x}$ images that determine whether we can give f arguments to k decimals.

8.2. FINDING WHERE A FUNCTION'S GRADIENT IS ZERO

The process described in the preceding section is slow, and the computer time required may increase at a more than linear rate as the number k of correct decimal places increases. Often a function f has one or more local extrema inside a certain region R, and one must locate these points at which the f gradient is $\mathbf{0}$. We assume that all f partial derivatives can be computed inside R.

This problem is similar to Problem 5.6 of finding in a given elementary region R the points where n functions of n variables are simultaneously zero. For our problem we have n partial derivative functions $\frac{\partial f}{\partial x_1}, \frac{\partial f}{\partial x_2}, \ldots, \frac{\partial f}{\partial x_n}$ defining the gradient of f, and we want the points where these functions are simultaneously zero. As with the zero finding problem, to get a solvable problem we must require the f gradient to be nonzero on the R boundary. Also, as with Problem 5.6, to avoid computerbound varieties of the problem, we require the f gradient to be zero only at isolated points in R.

For the gradient problem we search for points where the functions $\frac{\partial f}{\partial x_1}$, $\frac{\partial f}{\partial x_2}, \ldots, \frac{\partial f}{\partial x_n}$ are simultaneously zero. Previously we used $\mathbf{f}$ to denote the n component functions $f_1, f_2, \ldots, f_n$. Now we use the symbol $\frac{\partial f}{\partial \mathbf{x}}$ to denote the n partial derivatives $\frac{\partial f}{\partial x_1}, \frac{\partial f}{\partial x_2}, \ldots, \frac{\partial f}{\partial x_n}$. The Jacobian of $\mathbf{f}(\mathbf{x})$, used in dealing with the zero finding problems of Chapter 5, now becomes the Jacobian of $\frac{\partial f}{\partial \mathbf{x}}(\mathbf{x})$, which has the form

$$
\det \begin{bmatrix}
\dfrac{\partial^2 f}{\partial x_1^2}(\mathbf{x}) & \dfrac{\partial^2 f}{\partial x_1 \partial x_2}(\mathbf{x}) & \cdots & \dfrac{\partial^2 f}{\partial x_1 \partial x_n}(\mathbf{x}) \\[2ex]
\dfrac{\partial^2}{\partial x_2 \partial x_1}(\mathbf{x}) & \dfrac{\partial^2 f}{\partial x_2^2}(\mathbf{x}) & \cdots & \dfrac{\partial^2 f}{\partial x_2 \partial x_n}(\mathbf{x}) \\[2ex]
\vdots & \vdots & \vdots\vdots & \vdots \\[2ex]
\dfrac{\partial^2 f}{\partial x_n \partial x_1}(\mathbf{x}) & \dfrac{\partial^2 f}{\partial x_n \partial x_2}(\mathbf{x}) & \cdots & \dfrac{\partial^2 f}{\partial x_n^2}(\mathbf{x})
\end{bmatrix}
$$

This determinant is called the *Hessian* of f, and the corresponding matrix is the Hessian matrix.

We call a point $\mathbf{x}$ where the gradient is zero a simple zero-gradient-point if the Hessian of f at $\mathbf{x}$ is nonzero. Our problem, analogous to Problem 5.6, is

SOLVABLE PROBLEM 8.2: For any elementary function $f(\mathbf{x})$ with f and its gradient defined on an elementary region R, with the gradient being $\mathbf{0}$ only at isolated points in R, for a specific positive integer m, locate to m decimals a point on the boundary of R where all components of the gradient are less in magnitude than 10^{-m} and halt, or, if the gradient is not $\mathbf{0}$ on the boundary of R, bound all points in R where f has a zero gradient by giving to k decimals

1. points identified as simple zero-gradient-points, or
2. points identified as containing within their tilde-box a region where $\max_i |\frac{\partial f}{\partial x_i}(\mathbf{x})| < 10^{-k}$, the region identified as "containing at least one zero-

gradient-point" if the topological degree of $\frac{\partial f}{\partial \mathbf{x}}$ over a bounding box is nonzero.

A procedure for solving this problem is obtained by adapting the procedure used for Problem 5.6. That procedure requires at some stages computing the Jacobian of $\mathbf{f}$ over an $\mathbf{x}$ subbox $B^{(\mathbf{x})}$ derived from some container box $B^{(C)}$ of the unit box $B^{(\mathbf{u})}$. To get the Jacobian we obtain a degree 1 series expansion of each component function f_i over $B^{(\mathbf{x})}$. The coefficient of the f_i series term in $(x - x_j)$ yields the value for $\frac{\partial f_i}{\partial x_j}$ in the Jacobian. For Problem 8.2, the equivalent computation is finding the Hessian over $B^{(\mathbf{x})}$, and here we obtain a degree 2 series expansion of f. Line 4.39 indicates that one half the coefficient of the series term in $(x - x_i)^2$ yields the value for $\frac{\partial^2 f}{\partial x_i^2}$, and the coefficient of the series term in $(x - x_i)(x - x_j)$ yields the value for $\frac{\partial^2 f}{\partial x_i \partial x_j}$.

With Problem 8.2 it is possible to determine for each point $\mathbf{x}_0$ identified as a simple zero-gradient-point, whether $\mathbf{x}_0$ determines a local maximum, a local minimum, or a saddle point for f. To see this, let us take any such point $\mathbf{x}_0$ and translate $\mathbf{x}$ coordinates to make this point the origin. If we let H denote the Hessian matrix evaluated at $\mathbf{0}$, the Taylor series expansion of f about the series expansion point $\mathbf{0}$, up to degree 2 terms, may be expressed in matrix–vector notation as

$$f(\mathbf{x}) = f(\mathbf{0}) + \frac{1}{2} \mathbf{x}^T H \mathbf{x} + \cdots$$

The Hessian matrix H is symmetric, and accordingly there exists an orthogonal matrix P with inverse P^T, which transforms H to a real diagonal matrix J:

$$J = P^T H P$$

The eigenvalues of H are the diagonal elements of J. Because H equals PJP^T, and $(P^T\mathbf{x})^T = \mathbf{x}^T P$, the equation for $\mathbf{f}(\mathbf{x})$ can be written

$$f(\mathbf{x}) = f(\mathbf{0}) + \frac{1}{2} (P^T\mathbf{x})^T J (P^T\mathbf{x}) + \ldots$$

If all the eigenvalues of H are positive, then $f(\mathbf{x})$ equals $f(\mathbf{0})$ plus a sum of squares with positive coefficients, which indicates that at our zero-gradient-point $\mathbf{x}_0$, f has a local minimum. Similarly, if all the eigenvalues of H are negative, then f has a local maximum at $\mathbf{x}_0$, and if the eigenvalues are a mixture of positive and negative quantities, then f has a saddlepoint at $\mathbf{x}_0$. No eigenvalue of H can be zero because at a simple zero-gradient-point the determinant of H is nonzero.

In Section 5.5 a procedure involving the Jacobian matrix was described for determining whether a zero lies in a container box $B^{(C)}$. That procedure is readily adapted to the Hessian matrix to determine whether a zero-gradient-point lies in a box $B^{(x)}$. The process of clearing the matrix columns can be modified so that the signs of the Hessian matrix eigenvalues are determined. Then if a simple zero-gradient-point $\mathbf{x}_0$ is determined, its extremum type also is known. Instead of searching a matrix column for its best clearing element, we use the diagonal element whenever it does not contain 0. The first step is to use the interval element h_{11}, if it does not contain 0, to clear column 1. To clear the column 1 element h_{k1} from row k, we subtract row 1 multiplied by the interval quantity h_{k1}/h_{11}. The clearing of column 1 is equivalent to a premultiplication of H by a matrix Q identical to the identity matrix I except for the elements in column 1 below the diagonal, the element q_{k1} being $-h_{k1}/h_{11}$. After column 1 is cleared, the changed matrix is QH, which, except for the elements of row 1 to the right of h_{11}, is identical to the symmetric matrix QHQ^T. The matrix QHQ^T has its row 1 elements to the right of h_{11} all zero. This implies that the sign of the h_{11} interval is the sign of an eigenvalue of QHQ^T. We show below that the matrices H and QHQ^T have the same number of positive eigenvalues and the same number of negative eigenvalues. So the sign of h_{11} is the sign of an eigenvalue for any Hessian matrix evaluated at any point of $B^{(x)}$. The matrix QHQ^T and our matrix QH are identical below row 1, so now when we clear column 2, by reasoning as before, if the element h'_{22} does not contain 0, its sign gives the sign of another eigenvalue for a Hessian matrix evaluated at any point of $B^{(x)}$. Thus when we move from one column to the next, the matrix elements we treat are a submatrix of a symmetric matrix being gradually brought to diagonal form, and the eigenvalues of this matrix match in sign the eigenvalues of our starting matrix H. If any two diagonal elements obtained in the process of bringing H to upper triangular form are of opposite signs, then any simple zero gradient point in $B^{(x)}$ must be a saddle point. If all the diagonal elements are of the same sign, any simple zero gradient point is a local maximum or a local minimum according as the common sign is negative or positive.

To see that H and QHQ^T have the same eigenvalue sign structure, let Q_u be the matrix $(1 - u)I + uQ$ where u is a parameter varying in [0, 1]. Note that det $Q_u = 1$, so the varying matrix $Q_u H Q_u^T$ has a determinant that always equals det H. Also, this varying matrix is symmetric, so always has real eigenvalues, which match those of H when $u = 0$ and match those of QHQ^T when $u = 1$. Since the determinant of a matrix equals the product of the matrix eigenvalues, and we have been supposing that det H is nonzero, the varying eigenvalues of $QuHQ_u^T$ can never change sign by passing through zero.

It may happen in the process of bringing H to upper triangular form, that a diagonal element in a certain column contains 0, but there is an element lower down in the column that does not contain 0. By a row exchange, this element is brought into the diagonal position for the clearing operation. Whenever this occurs, this interrupts the matrix diagonalization process we have been describing. But because a symmetric matrix cannot have all its eigenvalues the same sign unless every two-square submatrix in diagonal position has a positive determinant, this state of affairs implies that a simple zero-gradient-point in $B^{(x)}$ is a saddle point.

EXERCISES

1. Call `maxmin` and obtain to five decimals the minimum of the function $5(y - x^2)^2 + (1 - x)^2$ in the square determined by the inequalities $-10 \le x \le 10$, $-10 \le y \le 10$. A three-dimensional graph of the function has a descending valley along the curve $y = x^2$ with the bottom of the valley at $(1, 1)$, where the minimum occurs.

2. Call `findextr` and obtain to five decimals the interior minimum point of the function of the preceding exercise in the square given. As the multiplier of $(y - x^2)^2$ is increased from its value of 5, the time to locate the minimum increases rapidly, and the problem easily becomes computerbound. If the multiplier is set at 100, the function is called Rosenbrock's function.

3. Call `findextr` and find to 10 decimals the local minima and local maxima of the function $\cos(2x + 1) + 2 \cos(3x + 2) + 3 \cos(4x + 3) + 4 \cos(5x + 4)$ in the interval $[-10, 10]$. There are 32 local extrema in $[-10, 10]$.

4. The book by Ratschek and Rokne [6] suggests the function

$$4x_1^2 - 2.1x_1^4 + x_1^6/3 + x_1x_2 - 4x_2^2 + 4x_2^4$$

to test mimimization programs. Call `maxmin` to find to five decimals the minimum of this function in the square defined by $-10 \le x_1 \le 10$, $-10 \le x_2 \le 10$.

5. Another test function given by Ratschek and Rokne is

$$\sqrt{1 - x_1^2}\cos x_3 + \frac{\sqrt{1 - x_2^2}}{1 + x_3^2} + 2x_3 e^{-x_1}$$

Call `maxmin` and find to five decimals the minimum of this function in the cube defined by $-1 \le x_1 \le 1$, $-1 \le x_2 \le 1$, $0 \le x_3 \le 10$.

NOTES

N8.1 Three general references that deal with optimization problems are by Fletcher [1], Hansen [2], and Ratschek and Rokne [6].

N8.2 Schaefer [7, 8] has treated optimization problems with range arithmetic.

REFERENCES

1. Fletcher, R., *Practical Methods of Optimization*, 2nd Edit., John Wiley & Sons, Chichester, 1987.

2. Hansen, E., *Global Optimization Using Interval Analysis*, Marcel Dekker, Inc., New York, 1992.

3. Ichida, K. and Fujii, Y., An interval arithmetic method for global optimization, *Computing* **23** (1979), 85–97.

4. Moore, R. E., *Interval Analysis, Prentice-Hall*, Englewood Cliffs, NJ, 1966.

5. Moore, R. E., On computing the range of values of a rational function of n variables over a bounded region, *Computing* **16** (1976), 1–15.

6. Ratschek, H. and Rokne, J., *New Computer Methods for Global Optimization*, Ellis Horwood and John Wiley, 1988.

7. Schaefer, M. J., Precise optimization using range arithmetic, *J. Computational and Appl. Math.* **53** (1994), 341–351.

8. Schaefer, M. J., Verification of constrained minima, *J. Computational and Appl. Math.* **67** (1996), 195–205.

9. Skelboe, S., Computation of rational interval functions, *BIT* **14** (1974), 87–95.

IX

NUMERICAL SOLUTION OF ORDINARY DIFFERENTIAL EQUATIONS

9.1. INTRODUCTION

A function $y(x)$ satisfies an *ordinary differential equation of the first order* if it satisfies an equation of the form

$$y' = f(x, y)$$

More generally, a function $y(x)$ satisfies an *ordinary differential equation of order n* if the n-th derivative of $y(x)$ is connected with lower order derivatives by the relation

$$y^{(n)} = f(x, y, y', \ldots, y^{(n-1)}) \tag{9.1}$$

Thus the equation

$$y'' = (y' + xy)^2 \tag{9.2}$$

is an ordinary differential equation of order 2. The differential equation 9.1 is *linear* if it can be expressed in the form

$$y^{(n)} + g_{n-1}(x)y^{(n-1)} + \ldots + g_1(x)y' + g_0(x)y = b(x)$$

Equation 9.2 is not a linear differential equation, but the equation below is.

$$y''' = xy'' - x^2y' + 3y - \sin(x)$$

More generally, s functions $y_1(x)$, $y_2(x)$, . . . , $y_s(x)$ satisfy s interrelated differential equations of respective orders $n_1, n_2, \ldots, n_s$ if the relations below are true:

$$
\begin{aligned}
y_1^{(n_1)} &= f_1(x; y_1, \ldots, y_1^{(n_1-1)}; y_2, \ldots, y_2^{(n_2-1)}; \ldots; y_s, \ldots, y_s^{(n_s-1)}) \\
y_2^{(n_2)} &= f_2(x; y_1, \ldots, y_1^{(n_1-1)}; y_2, \ldots, y_2^{(n_2-1)}; \ldots; y_s, \ldots, y_s^{(n_s-1)}) \\
&\vdots \qquad \vdots \\
y_s^{(n_s)} &= f_s(x; y_1, \ldots, y_1^{(n_1-1)}; y_2, \ldots, y_2^{(n_2-1)}; \ldots; y_s, \ldots, y_s^{(n_s-1)})
\end{aligned}
\qquad (9.3)
$$

The i-th differential equation is linear if it can be expressed in the form

$$y_i^{(n_i)} + \sum_{j=1}^{s} \sum_{k=0}^{n_j-1} g_{i,j,k}(x)y_j^{(k)} = h_i(x)$$

For example, the functions y_1 and y_2 may satisfy the interrelated equations

$$y_1'' = y_1^2 + y_1'y_2$$

$$y_2'' = 2y_1 + 3y_2$$

The second equation above, but not the first, is linear.

A set of interrelated differential equations is equivalent to a set of interrelated **first order** differential equations as shown below, involving the functions $u_1(x)$, $u_2(x)$, . . . , $u_m(x)$.

$$
\begin{aligned}
u_1' &= f_1(x, u_1, u_2, \ldots, u_m) \\
u_2' &= f_2(x, u_1, u_2, \ldots, u_m) \\
&\vdots \qquad \vdots \\
u_m' &= f_m(x, u_1, u_2, \ldots, u_m)
\end{aligned}
\qquad (9.4)
$$

Thus for the system consisting of the single equation 9.1, we may set $u_i(x)$ equal to the $(i-1)$-th derivative of $y(x)$, for $i = 1, 2, \ldots, n$, to obtain the equations

$$
\begin{aligned}
u_1' &= u_2 \\
u_2' &= u_3 \\
&\vdots \qquad \vdots \\
u_{n-1}' &= u_n \\
u_n' &= f(x, u_1, u_2, \ldots, u_n)
\end{aligned}
$$

Similarly, the system of equations 9.3 can be rewritten in terms of m variables

u_i, where $m = \Sigma_{i=1}^{s} n_i$. If $\mathbf{u}(x)$ denotes the vector with components $u_1(x)$, $u_2(x)$, . . . , $u_m(x)$, the system 9.4 may be written concisely as

$$\mathbf{u}' = \mathbf{f}(x, \mathbf{u}) \qquad (9.5)$$

Here $\mathbf{f}$ is an m-component vector function of x and $\mathbf{u}$. From now on, we use 9.5 with m unspecified to designate the system under study.

9.2. TWO STANDARD PROBLEMS OF ORDINARY DIFFERENTIAL EQUATIONS

In general, with any instance of the system 9.5, one finds that over a particular x interval $[a, b]$ there is a family of functions $\mathbf{u}(x)$ satisfying the equation. To obtain a unique solution, additional conditions are required.

The *initial value problem* for a system 9.5 is the determination of a solution $\mathbf{u}(x)$ in $[a, b]$ such that $\mathbf{u}$ equals a specified value $\mathbf{v}$ at a particular argument x_1 in $[a, b]$. Most often x_1 is at the left endpoint a of $[a, b]$, and we will assume that this is the case.

The *two-point boundary-value problem* for a system 9.5 is the determination of a solution $\mathbf{u}(x)$ in $[a, b]$ such that $\mathbf{u}(a)$ and $\mathbf{u}(b)$ satisfy m relations

$$g_i(\mathbf{u}(a), \mathbf{u}(b)) = 0 \qquad i = 1, 2, \ldots, m$$

where each function g_i has the $2m$ variables $u_1(a), \ldots, u_m(a), u_1(b), \ldots,$ $u_m(b)$. These relations are called the *boundary conditions*, and can be written in the vector form shown below if we understand that vectors always have m components, the same number of components as the system 9.5.

$$\mathbf{g}(\mathbf{u}(a), \mathbf{u}(b)) = \mathbf{0} \qquad (9.6)$$

The vector boundary condition (9.6) is *linear* if it can be expressed in the form

$$A\mathbf{u}(a) + B\mathbf{u}(b) = \mathbf{c}$$

where A and B are m-square real matrices, and $\mathbf{c}$ is an m-vector. For instance, if $A = I, B = -I$, and $\mathbf{c} = \mathbf{0}$, we have the *periodic* boundary condition

$$\mathbf{u}(a) = \mathbf{u}(b)$$

Linear boundary conditions are often encountered, and a frequent case is the *separated* variety:

$$A\mathbf{u}(a) = \mathbf{c}_a; \qquad B\mathbf{u}(b) = \mathbf{c}_b$$

Here m_a of the rows of A are zero rows, with the corresponding components of

c_a being zero, m_b of the rows of B are zero rows with the corresponding components of c_b being zero, and $m_a + m_b = m$. A particularly simple set of separated linear boundary conditions is given by the system of equations below.

$$u_{n_i}(a) = c_{n_i} \qquad i = 1, \ldots, k; \qquad u_{n_i}(b) = c_{n_i} \qquad i = k + 1, \ldots, m$$

Here m conditions are obtained by fixing in some fashion m of the $2m$ components of $\mathbf{u}(a)$ and $\mathbf{u}(b)$. Note that if all m components of $\mathbf{u}(a)$ are fixed, or all m components of $\mathbf{u}(b)$, then the two-point boundary-value problem becomes the initial value problem. Thus the two-point boundary-value problem includes the initial value problem as a special case.

Often a differential equation problem that appears to be different from either of the two basic types can be changed in some way to fit one of the patterns. For example, the motion of a weightless rigid pendulum of length 1, with its attached end at the origin and its free end with a unit mass at the point (x,y) of the vertical cartesian plane, is determined by the equations below, obtained by calculus of variations methods [2, p5]:

$$\frac{d^2x}{dt^2} = Kx; \qquad \frac{d^2y}{dt^2} = Ky - g; \qquad x^2 + y^2 = 1 \qquad (9.7)$$

The constant g equals the gravitational acceleration. Here a force along the pendulum has x and y components Kx and Ky where K is an unknown time dependent parameter. Suppose the values of x and y are desired when an initial velocity v_0 is imparted to the free end of the pendulum at its rest position. First the parameter K can be eliminated from the two differential equations to obtain

$$y \frac{d^2x}{dt^2} - x \frac{d^2y}{dt^2} = gx$$

The third equation of 9.7 can be differentiated twice with respect to t to obtain

$$x \frac{d^2x}{dt^2} + y \frac{d^2y}{dt^2} = -\left(\frac{dx}{dt}\right)^2 - \left(\frac{dy}{dt}\right)^2$$

When these two equations are solved for d^2x/dt^2 and d^2y/dt^2, we obtain the equations and initial conditions below:

$$\frac{d^2x}{dt^2} = gxy - x\left[\left(\frac{dx}{dt}\right)^2 + \left(\frac{dy}{dt}\right)^2\right]$$

$$\frac{d^2y}{dt^2} = -gx^2 - y\left[\left(\frac{dx}{dt}\right)^2 + \left(\frac{dy}{dt}\right)^2\right] \qquad (9.8)$$

$$x(0) = 0; \qquad \frac{dx(0)}{dt} = v_0; \qquad y(0) = -1; \qquad \frac{dy(0)}{dt} = 0$$

If we identify x, dx/dt, y, dy/dt, and t with the variables u_1, u_2, u_3, u_4, and x, we obtain the initial value problem

$$u_1' = u_2$$

$$u_2' = gu_1u_3 - u_1[u_2^2 + u_4^2]$$

$$u_3' = u_4$$

$$u_4' = -gu_1^2 - u_3[u_2^2 + u_4^2]$$

$$u_1(0) = 0; \qquad u_2(0) = v_0; \qquad u_3(0) = -1; \qquad u_4(0) = 0$$

As another example, consider the eigenvalue problem of a vibrating string:

$$y'' + \lambda y = 0$$

The eigenvalues λ are sought such that there is a solution function $y(x)$ in $[0, \pi]$ satisfying the boundary conditions

$$y(0) = y(\pi) = 0 \qquad \text{and} \qquad y'(0) = 1$$

The condition $y'(0) = 1$ is present to "normalize" the solution function. (The solution to this problem is $y = (1/\sqrt{\lambda}) \sin (\sqrt{\lambda} x)$, for $\lambda = 1^2, 2^2, 3^2, \ldots$)
Here if we identify u_1, u_2, and u_3 with y, y', and λ, we have a two-point boundary-value problem with the differential equations

$$u_1' = u_2$$

$$u_2' = u_3u_1$$

$$u_3' = 0$$

and the separated boundary conditions

$$u_1(0) = 0$$

$$u_2(0) = 1$$

$$u_1(\pi) = 0$$

As a final example, consider the Van der Pol "free boundary problem" [6, p76]:

$$u'' + (u^2 - 1)u' + u = 0 \qquad u(0) = -u(T) \qquad \text{and} \qquad u'(0) = u'(T) = 0$$

Here the independent variable x is time and the coordinate T, the half period, is unknown and is to be determined. We can convert the problem to one with fixed boundary points, if we set x equal to $T\hat{x}$, thinking of $\hat{x}$ as a variable in $[0, 1]$. With this change in the independent variable, we obtain the equations

$$\frac{u''}{T^2} + (u^2 - 1)\frac{u'}{T} + u = 0 \qquad u(0) = -u(1) \qquad \text{and} \qquad u'(0) = u'(1)$$

The parameter T now can be treated like the parameter λ in the preceding example and be replaced by a variable u_3 with zero derivative. The differential equations then are

$$u_1' = u_2$$

$$u_2' = -[(u_1^2 - 1)u_2 u_3 + u_1 u_3^2]$$

$$u_3' = 0$$

9.3. A SOLVABLE INITIAL VALUE PROBLEM

There are certain difficulties that may be encountered in treating the initial value problem numerically. Let us consider the simple case where m, the number of functions, is 1, so our problem is to find a solution to

$$u_1' = f_1(x, u_1)$$

$$u_1(a) = v_1$$

Sometimes a solution in $[a, b]$ is not possible because the solution function $u_1(x)$ tends to infinity at some point within the designated interval $[a, b]$. For instance, suppose in the x interval $[0, 2]$ we wanted a solution to the problem

$$u_1' = u_1^2$$

$$u_1'(0) = 1$$

The *nonlinear* differential equation is separable and the solution is easily found to be

$$u_1(x) = \frac{1}{1 - x}$$

Because $\lim\limits_{x \to 1^-} u_1, = \infty$, an accurate numerical solution can not reach the midpoint of the $[0, 2]$ interval. Thus we need assurance that the solution $\mathbf{u}(x)$ we are approximating has bounded components in $[a, b]$.

We will suppose that for the initial value problem $\mathbf{u}' = \mathbf{f}(x, \mathbf{u})$, $\mathbf{u}(a) = \mathbf{v}$, we can restrict the solution to a *containment box* $B_{[a,b],M}$ of $m + 1$ dimensions, defined by the relations

$$a \le x \le b; \qquad u_i = v_i \pm M \qquad i = 1, \ldots, m$$

We require that the solution never touches a side of the box, except of course the sides defined by $x = a$ or $x = b$. For instance, if we arbitrarily choose a large positive constant M to define a box $B_{[a,b],M}$, and then verify that in the box the components of $|\mathbf{f}(x,\mathbf{u})|$ are less than $M/(b-a)$, we have a containment box, because $|u_i'| < M/(b-a)$ for x in $[a,\ b]$ implies $|u_i(x) - v_i| < M$ for x in $[a,\ b]$.

A difficulty of a different variety is illustrated by the problem of finding in $[0,1]$ a solution to the problem

$$u_1' = u_1^{1/3}$$
$$u_1(0) = 0 \tag{9.9}$$

The differential equation, again nonlinear, is separable, and it is not difficult to verify that there are an infinite number of solutions to this problem. We use a parameter t to distinguish among the various solutions:

$$u_1(x;\ t) = \begin{cases} 0 & \text{for } 0 \le x \le t \\ \left[\dfrac{2}{3}(x - t)\right]^{3/2} & \text{for } t > x \le 1 \end{cases}$$

The parameter t may assume any value in $[0,\ 1]$. When $t = 1$, the solution is identically zero; when $t = 0$, the solution is $[\frac{2}{3}x]^{\frac{3}{2}}$; and when t is between 0 and 1, the solution combines features of the two extreme solutions. Note that when t is in $(0,1)$, the two parts of the solution join without introducing a discontinuity in u_1 or its derivative.

A restriction on the function $\mathbf{f}(x,\mathbf{u})$ of 9.5 that eliminates the possibility of such multiple solutions is the *Lipschitz condition*.

DEFINITION 9.1: The elementary function $\mathbf{f}(x,\mathbf{u})$ satisfies the Lipschitz condition in a containment box $B_{[a,b],M}$ if there is a positive constant l such that for any two arguments $(x,\ \mathbf{u}^{(\alpha)})$ and $(x,\ \mathbf{u}^{(\beta)})$ in $B_{[a,b],M}$ that share the same x coordinate, we have

$$|f_i(x,\ \mathbf{u}^{(\alpha)}) - f_i(x,\ \mathbf{u}^{(\beta)})| \le l \sum_{j=1}^{m} |u_j^{(\alpha)} - u_j^{(\beta)}| \qquad i = 1,\ldots,m$$

Thus if m equals 3, the following inequalities are required:

$$|f_1(x,\ \mathbf{u}^{(\alpha)}) - f_1(x,\ \mathbf{u}^{(\beta)})| \le l(|u_1^{(\alpha)} - u_1^{(\beta)}| + |u_2^{(\alpha)} - u_2^{(\beta)}| + |u_3^{(\alpha)} - u_3^{(\beta)}|)$$
$$|f_2(x,\ \mathbf{u}^{(\alpha)}) - f_2(x,\ \mathbf{u}^{(\beta)})| \le l(|u_1^{(\alpha)} - u_1^{(\beta)}| + |u_2^{(\alpha)} - u_2^{(\beta)}| + |u_3^{(\alpha)} - u_3^{(\beta)}|)$$
$$|f_3(x,\ \mathbf{u}^{(\alpha)}) - f_3(x,\ \mathbf{u}^{(\beta)})| \le l(|u_1^{(\alpha)} - u_1^{(\beta)}| + |u_2^{(\alpha)} - u_2^{(\beta)}| + |u_3^{(\alpha)} - u_3^{(\beta)}|)$$

If m equals 1, the requirement is

$$|f_1(x,\ u_1^{(\alpha)}) - f_1(x,\ u_1^{(\beta)})| \le l|u_1^{(\alpha)} - u_1^{(\beta)}|$$

Note that it is impossible to find a Lipschitz constant l for the one component function $\mathbf{f}$ of 9.9 over any containment box that contains a segment of the line $u_1 = 0$, because

$$\frac{|f_1(x, u_1) - f_1(x, 0)|}{|u_1 - 0|} = \frac{|u_1^{1/3}|}{|u_1|} = |u_1|^{-2/3}$$

and $|u_1|^{-\frac{2}{3}}$ tends to infinity as u_1 approaches zero.

We now have the elements for

SOLVABLE PROBLEM 9.1: Given the initial value problem $\mathbf{u}' = \mathbf{f}(x, \mathbf{u})$, $\mathbf{u}(a) = \mathbf{v}$, where $\mathbf{f}(x, \mathbf{u})$ is an elementary function defined in a containment box $B_{[a,b],M}$ for which there is the Lipschitz constant l, for any argument c in $[a, b]$, find to k decimals $\mathbf{u}(c)$.

We prove the problem is solvable and the solution unique by showing:

1. For a small enough positive number ϵ it is possible to construct in $[a,b]$ a function $\mathbf{u}^{(\epsilon)}(x)$ that remains in the containment box $B_{[a,b],M}$, equals $\mathbf{v}$ at $x = a$, and in $[a,b]$ satisfies

$$\left| \frac{d}{dx} u_i^{(\epsilon)}(x) - f_i(x, \mathbf{u}^{(\epsilon)}) \right| \leq \epsilon \qquad i = 1, \dots, m \qquad (9.10)$$

2. For any two functions $\mathbf{u}^{(\epsilon_1)}(x)$ and $\mathbf{u}^{(\epsilon_2)}(x)$ contained within $B_{[a,b],M}$, satisfying 9.10 in $[a,b]$ and equaling $\mathbf{v}$ at $x = a$, allowing even exact solution cases $\epsilon_1 = 0$ or $\epsilon_2 = 0$, we have for x in $[a,b]$

$$\left| u_i^{(\epsilon_1)}(x) - u_i^{(\epsilon_2)}(x) \right| \leq \frac{\epsilon_1 + \epsilon_2}{ml} (e^{ml(x-a)} - 1) \qquad i = 1, \dots, m \quad (9.11)$$

There cannot be two distinct solutions to the initial value problem, because if these two solutions are $\mathbf{u}^{(\epsilon_1)}$ and $\mathbf{u}^{(\epsilon_2)}$ with $\epsilon_1 = \epsilon_2 = 0$, 9.11 implies that the solutions are identical. If $\mathbf{u}(x)$ is the unique solution to the problem, then 9.11 implies

$$\left| u_i^{(\epsilon)}(x) - u_i(x) \right| \leq \frac{\epsilon}{ml} (e^{ml(x-a)} - 1) \qquad i = 1, \dots, m \qquad (9.12)$$

Thus we can approximate $\mathbf{u}(c)$ as closely as we please by $\mathbf{u}^{(\epsilon)}(c)$ with ϵ chosen sufficiently small. We need to show how to obtain a function $\mathbf{u}^{(\epsilon)}(x)$ and we need to prove 9.11.

The construction of the function $\mathbf{u}^{(\epsilon)}(x)$ is done in steps. The k-th step begins with the value $\mathbf{u}^{(\epsilon)}(x_k)$ found by the previous step and extends $\mathbf{u}^{(\epsilon)}(x)$ into an interval $[x_k, x_{k+1}]$. (In the first step x_1 equals a and $\mathbf{u}^{(\epsilon)}(x_1)$ equals $\mathbf{v}$.) To fix the next endpoint x_{k+1}, we do a succession of interval evaluations of $\mathbf{f}(x,\mathbf{u})$ components over small containment boxes, setting the primitives according to the plan

$$x = x_k \pm 2^{-q}, \qquad u_i = u_i^{(\epsilon)}(x_k) \pm 2^{-q}$$

This computation is done for $q = 1, 2, \ldots$, obtaining the succession of component intervals $f_i(x_k, \mathbf{u}^{(\epsilon)}(x_k)) \pm w_i^{(q)}$ until finally for $q = q_0$ we find $w_i^{(q_0)} < \epsilon$ for all i. Then we take x_{k+1} equal to $x_k + 2^{-q_0}$, $m_i^{(\kappa)}$ equal to $f_i(x_k, \mathbf{u}^{(\epsilon)}(x_k))$, and in the interval $[x_k, x_{k+1}]$ set

$$u_i^{(\epsilon)}(x) = u_i^{(\epsilon)}(x_k) + m_i^{(k)}(x - x_k)$$

In this way we obtain piecewise linear solution components $u_i^{(\epsilon)}(x)$ reaching from $x = a$ to $x = b$. It is possible that the approximation exits the containment box at some step of our construction. In this case, the attempt to obtain an approximation satisfying 9.10 has failed. However, inequality 9.12 shows that this difficulty does not occur if ϵ is small enough.

It is true that our function $\mathbf{u}^{(\epsilon)}(x)$ fails to have a derivative at the interior division points x_k, because two different linear elements join at these points; but, this difficulty can be eliminated by adding to the expression above for $u_i^{(\epsilon)}(x)$ quadratic and cubic terms that do not change the value of $u_i^{(\epsilon)}(x)$ at the endpoints of $[x_k, x_{k+1}]$, but make its derivative vary over this interval and match the value $m_i^{(k+1)}$ at x_{k+1}, yet remain within the f_i interval:

$$u_i^{(\epsilon)}(x) = u_i^{(\epsilon)}(x_k) + m_i^{(k)}(x - x_k)$$
$$+ (m_i^{(k)} - m_i^{(k+1)})\left(\frac{(x - x_k)^2}{x_{k+1} - x_k} - \frac{(x - x_k)^3}{(x_{k+1} - x_k)^2}\right)$$

The inequality 9.11 is proved using the Lipschitz condition. We prove the more general result given below, where the two functions $\mathbf{u}^{(\epsilon_1)}(x)$ and $\mathbf{u}^{(\epsilon_2)}(x)$ no longer are required to equal $\mathbf{v}$ at $x = a$. The quantity δ is $\max_i |u_i^{(\epsilon_1)}(a) - u_i^{(\epsilon_2)}(a)|$.

$$|u_i^{(\epsilon_1)}(x) - u_i^{(\epsilon_2)}(x)| \le \delta e^{ml(x-a)} + \frac{\epsilon_1 + \epsilon_2}{ml}(e^{ml(x-a)} - 1) \quad i = 1, \ldots, m \qquad (9.13)$$

We have

$$\left|\frac{d}{dx} u_i^{(\epsilon_1)}(x) - \frac{d}{dx} u_i^{(\epsilon_2)}(x)\right| \le |f_i(x, \mathbf{u}^{(\epsilon_1)}(x)) - f_i(x, \mathbf{u}^{(\epsilon_2)}(x))| + \epsilon_1 + \epsilon_2$$

$$\le l \sum_j |u_j^{(\epsilon_1)}(x) - u_j^{(\epsilon_2)}(x)| + \epsilon_1 + \epsilon_2 \qquad (9.14)$$

Also because

$$u_i^{(\epsilon_1)}(x) - u_i^{(\epsilon_2)}(x) = u_i^{(\epsilon_1)}(a) - u_i^{(\epsilon_2)}(a) + \int_a^x \frac{d}{dt} (u_i^{(\epsilon_1)}(t) - u_i^{(\epsilon_2)}(t)) \, dt$$

we have

$$\left| u_i^{(\epsilon_1)}(x) - u_i^{(\epsilon_2)}(x) \right| \le \delta + \int_a^x \left| \frac{d}{dt} (u_i^{(\epsilon_1)}(t) - u_i^{(\epsilon_2)}(t)) \right| dt$$

Employing the inequality 9.14, we get

$$\left| u_i^{(\epsilon_1)}(x) - u_i^{(\epsilon_2)}(x) \right| \le \delta + l \int_a^x \sum_j |u_j^{(\epsilon_1)}(t) - u_j^{(\epsilon_2)}(t)| \, dt + (\epsilon_1 + \epsilon_2)(x - a) \qquad (9.15)$$

Now set $h(x)$ equal to $\int_a^x \sum_j | u_j^{(\epsilon_1)}(t) - u_j^{(\epsilon_2)}(t) | \, dt$. After summing the m component inequalities 9.15, we obtain

$$h'(x) - mlh(x) \le m[\delta + (\epsilon_1 + \epsilon_2)(x - a)]$$

Multiply this inequality by the positive quantity $e^{-ml(x-a)}$ to get

$$\frac{d}{dx} (h(x)e^{-ml(x-a)}) \le m[\delta + (\epsilon_1 + \epsilon_2)(x - a)]e^{-ml(x-a)}$$

We have then

$$h(x)e^{-ml(x-a)} = \int_a^x \frac{d}{dt} h(t)e^{-ml(t-a)} \, dt$$

$$\le \int_a^x m[\delta + (\epsilon_1 + \epsilon_2)(t - a)]e^{-ml(t-a)} \, dt$$

For the two parts of the definite integral on the right above, we obtain

$$\int_a^x m\delta e^{-ml(t-a)} \, dt = \frac{\delta}{-l} e^{-ml(t-a)} \Big|_a^x$$

$$= \frac{\delta}{l}(1 - e^{-ml(x-a)})$$

$$\int_a^x m(\epsilon_1 + \epsilon_2)(t - a)e^{-ml(t-a)} \, dt = m(\epsilon_1 + \epsilon_2)\left[(t - a)\frac{e^{-ml(t-a)}}{-ml} - \frac{e^{-ml(t-a)}}{m^2 l^2} \right]_a^x$$

$$= (\epsilon_1 + \epsilon_2)\left[(x - a)\frac{e^{-ml(x-a)}}{-l} + \frac{1 - e^{-ml(x-a)}}{ml^2} \right]$$

Hence

$$h(x)e^{-ml(x-a)} \leq \left(\frac{\delta}{l} + \frac{\epsilon_1 + \epsilon_2}{ml^2}\right)(1 - e^{-ml(x-a)}) - \frac{(\epsilon_1 + \epsilon_2)(x - a)}{l}e^{-ml(x-a)}$$

$$h(x) \leq \frac{\delta}{l}e^{ml(x-a)} + \frac{\epsilon_1 + \epsilon_2}{ml^2}(e^{ml(x-a)} - 1) - \frac{\delta}{l} - \frac{(\epsilon_1 + \epsilon_2)(x - a)}{l}$$

If we substitute for $h(x)$ in inequality 9.15, we get finally the inequality 9.13.

Now that Problem 9.1 has been shown to be solvable, we can consider making adjustments to it so that the required conditions are easier to verify. A more convenient problem is

SOLVABLE PROBLEM 9.2: Given the initial value problem $\mathbf{u}' = \mathbf{f}(x, \mathbf{u})$, $\mathbf{u}(a) = \mathbf{v}$, where $\mathbf{f}(x, \mathbf{u})$ is an elementary function differentiable with respect to its $\mathbf{u}$ variables over a containment box $B_{[a,b],M}$, for any argument c in $[a, b]$, find to k decimals $\mathbf{u}(c)$.

We no longer need require a Lipschitz constant l, because such a constant is implied by the differentiability condition. We can use Theorem 5.1 to show this. For any two points $(x, \mathbf{u}^{(\alpha)})$ and $(x, \mathbf{u}^{(\beta)})$ in the containment box, we have

$$|f_i(x, \mathbf{u}^{(\alpha)}) - f_i(x, \mathbf{u}^{(\beta)})| \leq \sum_{j=1}^{m} \left|\frac{\partial f_i(\mathbf{c}_i)}{\partial u_j}\right| |u_j^{(\alpha)} - u_j^{(\beta)}|$$

In the containment box, any partial derivative $\partial f_i/\partial u_j$ is bounded in magnitude, and a bound large enough to serve for all these derivatives could be taken as l.

Although Problem 9.2 specifies a containment box to be certain that the solution is defined throughout $[a, b]$, it is possible to dispense with this requirement if an escape is allowed. We still need to specify a region within which the differentiability condition applies, but the region can depend on x only, allowing all components of $\mathbf{u}$ to range in $(-\infty, +\infty)$:

SOLVABLE PROBLEM 9.3: Given the initial value problem $\mathbf{u}' = \mathbf{f}(x, \mathbf{u})$, $\mathbf{u}(a) = \mathbf{v}$, where $\mathbf{f}(x, \mathbf{u})$ is an elementary function differentiable with respect to its $\mathbf{u}$ variables at any argument $(x, \mathbf{u})$ with x in $[a, b]$, for any number c in $[a, b]$, find to k decimals $\mathbf{u}(c)$, or else indicate that at some point in (a, c) a component of $\mathbf{u}(x)$ becomes larger in magnitude than a prescribed positive bound M.

9.4. LINEAR DIFFERENTIAL EQUATIONS

Suppose for an initial value problem all the individual differential equations of $\mathbf{u}' = \mathbf{f}(x, \mathbf{u})$ are linear, that is, each equation may be written in the form

$$u_i' + \sum_{j=1}^{m} g_{ij}(x)u_j = h_i(x)$$

The function $\mathbf{f}(x, \mathbf{u})$ is then called *linear*. If all elementary functions $g_{i,j}$ and h_i are defined in the interval $[a, b]$ of the initial value problem, then all the solution components are bounded, and a containment box can be found. For we can find a positive constant p bounding the magnitude of all these elementary functions in $[a, b]$. We have then

$$|u_i'| \leq \sum_{j=1}^{m} |g_{ij}(x)||u_j| + |h_i(x)|$$

$$\leq \sum_{j=1}^{m} p|u_j| + p$$

Thus the magnitude of the components of $\mathbf{u}(x)$ can grow no faster than the components of the solution to the initial value problem

$$\mathbf{u}' = D\mathbf{u} + \mathbf{h}, \qquad \mathbf{u}(a) = |\mathbf{v}|$$

Here all elements of the matrix D and the components of the vector $\mathbf{h}$ equal the bound p. The solution to the equations above may grow rapidly but is never unbounded. Thus all the requirements of Problem 9.2 are met.

SOLVABLE PROBLEM 9.4: Given the initial value problem $\mathbf{u}' = \mathbf{f}(x, \mathbf{u})$, $\mathbf{u}(a) = \mathbf{v}$, where $\mathbf{f}(x, \mathbf{u})$ is a linear elementary function defined for x in $[a, b]$, find to k decimals $\mathbf{u}(c)$, where c is any x value in $[a, b]$.

9.5. SOLVING THE INITIAL VALUE PROBLEM BY POWER SERIES

In Chapter 4 we computed definite integrals of elementary functions using power series. Power series methods are useful also for solving differential equations. However to apply such methods, the function $\mathbf{f}(x, \mathbf{u})$ must be differentiable with respect to its variables to arbitrary order.

If $\mathbf{u}$ is the solution to Problem 9.2, for each component u_i, we have the differential equation $u_i' = f_i(x, u_1, \ldots, u_m)$. This equation can be differentiated any number of times with respect to x, because all partial derivatives of all functions f_i exist, so at any point x in $[a, b]$ each component of the solution $\mathbf{u}$ is infinitely differentiable. These components can be approximated by Taylor polynomials over small subintervals of $[a, b]$.

There is some similarity to the power series method of computing definite integrals described in Chapter 4, but the procedure here is more complicated. Suppose we have advanced our $\mathbf{u}(x)$ approximation from $x = a$ up to $x = x_k$

and have obtained the interval $\mathbf{m}^{(k)} \pm \mathbf{w}^{(k)}$ for $\mathbf{u}(x_k)$. Here we are using vector notation to represent the m components. The wid vector $\mathbf{w}^{(k)}$ is the *global error* for $\mathbf{u}$ at x_k. (In the first step of the process, $x_1 = a$, and $\mathbf{m}^{(1)} \pm \mathbf{w}^{(1)} = \mathbf{v} \pm \mathbf{0}$.) At the series expansion point $(x_k, \mathbf{m}^{(k)})$ we try to obtain a power series for the solution $\mathbf{u}(x)$ that can be used to approximate the solution at any x in the interval $[x_k, x_k + h]$. The step width h is adjusted so that various requirements to be described are satisfied. Generally the h used successfully in the preceding interval $[x_{k-1}, x_k]$ is tried first, but occasionally this h is doubled to test whether conditions have changed to allow a larger step width. If any of the required conditions are not met, h is halved, and the attempt repeated. In the starting x interval $[a, a + h]$, the first trial h is sizeable, say 0.5.

The series for u in powers of $x - x_k$ is found up to the terms in $(x - x_k)^n$ where n is reasonably large, say 12 or higher. The computation for the terms of $\mathbf{u}$ are made by a recursive procedure. Initially variables are set to

$$x = (x_k \pm h) + (1 \pm 0)(x - x_k)\bullet$$

$$\mathbf{u} = (\mathbf{m}^{(k)} \pm \hat{\mathbf{w}})\bullet$$

The wid vector $\hat{\mathbf{w}}$ has components larger than or identical to corresponding components of the global error vector $\mathbf{w}^{(k)}$. The vector $\hat{\mathbf{w}}$ must be chosen so that for any x in $[x_k, x_k + h]$ it is certain that $\mathbf{u}(x)$ is inside the containment box defined by $x = x_k \pm h$, $\mathbf{u} = \mathbf{m}^{(k)} \pm \hat{\mathbf{w}}$.

For the chosen step width h, whether or not a vector $\hat{\mathbf{w}}$ is allowable is determined by the outcome of a degree 0 series evaluation for $\mathbf{f}$:

$$\mathbf{f} = (\mathbf{f}_0 \pm \hat{\mathbf{w}}_0)\bullet$$

The containment condition is satisfied if

$$\mathbf{w}^{(k)} + h(|\mathbf{f}_0| + \hat{\mathbf{w}}_0) < \hat{\mathbf{w}} \tag{9.16}$$

A value for $\hat{\mathbf{w}}$ is desired with as small components as possible. A simple method of determining a suitable $\hat{\mathbf{w}}$ is to set $\hat{\mathbf{w}}$ initially to $\mathbf{w}^{(k)}$, evaluate $\mathbf{f}$, and if the inequality 9.16 fails for component j, then component j of $\hat{\mathbf{w}}$ is increased to the j component of the left side of 9.16 times a factor slightly above 1, say 1.1. The test 9.16 then is repeated after $\mathbf{f}$ is recalculated. After a fixed number of these attempts to adjust $\hat{\mathbf{w}}$ fail, h is halved, and the whole process restarted.

After $\hat{\mathbf{w}}$ is found, the next step is to determine all the terms of the $\mathbf{u}$ series. Because $\mathbf{u}' = \mathbf{f}$, a term $(\mathbf{f}_k \pm \mathbf{w}_k)(x - x_k)^k$ determined for the $\mathbf{f}$ series becomes the term $(\frac{1}{k+1}\mathbf{m}_k \pm \frac{1}{k+1}\mathbf{w}_k)(x - x_k)^{k+1}$ for $\mathbf{u}$. So the degree 0 term for $\mathbf{f}$ allows $\mathbf{u}$ to be changed to

$$\mathbf{u} = (\mathbf{m}^{(k)} \pm \hat{\mathbf{w}}) + (\mathbf{f}_0 \pm \hat{\mathbf{w}}_0)(x - x_k)\bullet$$

The computation for the **f** series now is carried further to obtain the degree 1 term of **f**, which is used to fix the degree 2 term of **u**. This cyclic procedure, of computing the degree j term of **f** and using it to fix the degree $j + 1$ term of the **u** series, continues until all the required terms of **u** are obtained.

If the global error vector $\mathbf{w}^{(k)}$ for $\mathbf{u}(x_k)$ were **0**, then by the series remainder formula 4.23, we would have

$$\mathbf{u}(x_k + h) = \mathbf{m}^{(k)} + h\mathbf{f}_0 + \frac{h^2}{2}\mathbf{f}_1 + \cdots + \frac{h^{n-1}}{n-1}\mathbf{f}_{n-2} + \frac{h^n}{n}(\mathbf{f}_{n-1} \pm \hat{\mathbf{w}}_{n-1})$$

The vector $\frac{h^n}{n}\hat{\mathbf{w}}_{n-1}$ is the *local error* over the interval $[x_k, x_k + h]$. The local errors accumulate and add to the global error as the solution is constructed from a toward b. If the rate of growth of global error due to the local error, $\frac{h^{n-1}}{n}\hat{\mathbf{w}}_{n-1}$, exceeds (in any component) a preset bound (which would depend on k, the number of correct decimals wanted), then h is halved, and the whole process restarted. The global error $\mathbf{w}^{(k+1)}$ equals the local error over $[x_k, x_k + h]$ plus a wid vector bounding the growth of $\mathbf{w}^{(k)}$ over $[x_k, x_k + h]$, the *global error carryover*. To obtain this vector, all **u** series terms of degree less than n must be assigned a wid to reflect the global error $\mathbf{w}^{(k)}$, and this requires a new computation of these terms. For this second calculation, x is fixed at x_k, because all these series terms are evaluated at $x = x_k$. The variables are set initially to

$$x = (x_k \pm 0) + (1 \pm 0)(x - x_k)\bullet$$

$$\mathbf{u} = (\mathbf{m}^{(k)} \pm \mathbf{w}^{(k)})\bullet$$

and the cyclic process described previously is repeated, and ends when the **u** term of degree $n - 1$ is obtained. The degree 1 term given above for the variable x is needed to make the computation yield the same mids as before but with wids changed. We distinguish these new **f** wids by using **w** in place of $\hat{\mathbf{w}}$, and have now

$$\mathbf{u}(x_k + h) = (\mathbf{m}^{(k)} \pm \mathbf{w}^{(k)}) + h(\mathbf{f}_0 \pm \mathbf{w}_0) + \frac{h^2}{2}(\mathbf{f}_1 \pm \mathbf{w}_1) + \cdots + \frac{h^n}{n}(\mathbf{f}_{n-1} \pm \hat{\mathbf{w}}_{n-1}) \quad (9.17)$$

Because $\mathbf{u}(x_k + h) = \mathbf{m}^{(k+1)} \pm \mathbf{w}^{(k+1)}$, we can write

$$\mathbf{m}^{(k+1)} = \mathbf{m}^{(k)} + h\mathbf{f}_0 + \frac{h^2}{2}\mathbf{f}_1 + \cdots + \frac{h^n}{n}\mathbf{f}_{n-1} \quad (9.18)$$

$$\mathbf{w}^{(k+1)} = \mathbf{w}^{(k)} + h\mathbf{w}_0 + \frac{h^2}{2}\mathbf{w}_1 + \cdots + \frac{h^{n-1}}{n-1}\mathbf{w}_{n-2} + \frac{h^n}{n}\hat{\mathbf{w}}_{n-1} \quad (9.19)$$

The global error carryover is

$$\mathbf{w}^{(k)} + h\mathbf{w}_0 + \frac{h^2}{2}\,\mathbf{w}_1 + \cdots + \frac{h^{n-1}}{n-1}\,\mathbf{w}_{n-2} \tag{9.20}$$

We can calculate $\mathbf{u}(x)$ for any x in $[x_k, x_k + h]$ by using (9.18) and (9.19) with h replaced by the value of $x - x_k$.

Let us apply the proposed system to the test case:

$$\mathbf{u}' = -\mathbf{u}$$

$$\mathbf{u}(0) = \mathbf{v} \tag{9.21}$$

The number of components m can be any positive integer, and the solution is $\mathbf{u} = e^{-x}\mathbf{v}$. Suppose we have carried the numerical solution up to the point $x = x_k$, obtaining $\mathbf{u}(x_k) = \mathbf{m}^{(k)} \pm \mathbf{w}^{(k)}$. Because $\mathbf{f} = -\mathbf{u}$, the $\mathbf{u}$ series term in $(x - x_k)^j$ has the coefficient $(-1)^j \frac{1}{j!}(\mathbf{m}^{(k)} \pm \mathbf{w}^{(k)})$. Therefore we obtain for $\mathbf{u}(x_k + h) = \mathbf{m}^{(k+1)} \pm \mathbf{w}^{(k+1)}$ the value

$$\mathbf{m}^{(k+1)} = \mathbf{m}^{(k)} - h\mathbf{m}^{(k)} + \frac{h^2}{2!}\,\mathbf{m}^{(k)} - \frac{h^3}{3!}\,\mathbf{m}^{(k)} + \cdots + (-1)^n\frac{h^n}{n!}\,\mathbf{m}^{(k)}$$

$$\mathbf{w}^{(k+1)} = \mathbf{w}^{(k)} + h\mathbf{w}^{(k)} + \frac{h^2}{2!}\,\mathbf{w}^{(k)} + \frac{h^3}{3!}\,\mathbf{w}^{(k)} + \cdots + \frac{h^{n-1}}{(n-1)!}\,\mathbf{w}^{(k)} + \frac{h^n}{n!}\,\hat{\mathbf{w}}$$

(By the rules of interval arithmetic, $c(\mathbf{m} \pm \mathbf{w}) = c\mathbf{m} \pm |c|\mathbf{w}$.) We see that $\mathbf{m}^{(k+1)}$ is nearly $e^{-h}\mathbf{m}^{(k)}$, a satisfactory result, but the global error carryover is nearly $e^h\mathbf{w}^{(k)}$. It is not possible to compute $\mathbf{u}(x)$ to k decimals for a large interval $[a, b]$, with the global error growing exponentially this way. It is clear that an improvement in the computation of $\mathbf{w}^{(k+1)}$ is needed.

9.6. AN IMPROVED GLOBAL ERROR

By using the type of interval arithmetic introduced in Section 7.4, the global error carryover computation can be greatly improved. The computation 9.20 is a wid computation depending only on the $\mathbf{u}$ intervals. To obtain an improved wid, we set $\mathbf{u}$ components to

$$u_i = m_i^{(k)} \pm w_i^{(k)} + (1 \pm 0)(u_i - m_i^{(k)})\bullet \qquad i = 1, \ldots, m$$

or, using matrix–vector notation, to

$$\mathbf{u} = \mathbf{m}^{(k)} \pm \mathbf{w}^{(k)} + (I \pm O)(\mathbf{u} - \mathbf{m}^{(k)})\bullet \tag{9.22}$$

All $\mathbf{f}$ computations are now made to degree 1 with respect to $\mathbf{u}$, so this changes the coefficient of the $\mathbf{f}$ series term in $(x - x_k)^j$ from $\mathbf{f}_j \pm \mathbf{w}_j$ to

$$\mathbf{f}_j \pm \mathbf{w}_j + (A_j \pm B_j)(\mathbf{u} - \mathbf{m}^{(k)})$$

where A_j and B_j are matrices. Instead of the expression 9.17 for $\mathbf{u}(x_k + h)$, we now obtain an expression that is degree 1 with respect to $\mathbf{u}$, with a degree 0 part identical to the right side of 9.17 and with a degree 1 part equal to

$$\left[(I \pm O) + h(A_0 \pm B_0) + \frac{h^2}{2}(A_1 \pm B_1) + \cdots + \frac{h^{n-1}}{n-1}(A_{n-2} \pm B_{n-2}) \right](\mathbf{u} - \mathbf{m}^{(k)})$$

$$= (A \pm B)(\mathbf{u} - \mathbf{m}^{(k)}) \tag{9.23}$$

where the matrices A and B are determined by the equations

$$A = I + hA_0 + \frac{h^2}{2}A_1 + \cdots + \frac{h^{n-1}}{n-1}A_{n-2}$$

$$B = hB_0 + \frac{h^2}{2}B_1 + \cdots + \frac{h^{n-1}}{n-1}B_{n-2}$$

We replace the global error carryover 9.20, which is degree 0 with respect to $\mathbf{u}$, with the degree 1 wid vector $(|A| + B)\mathbf{w}^{(k)}$. There is no change in the mid value 9.18 for $\mathbf{m}^{(k+1)}$, and also no change in the local error contribution to $\mathbf{w}^{(k+1)}$.

For the test problem 9.21 we find the coefficient of the $\mathbf{u}$ series term in $(x - x_k)^j$ equals

$$(-1)^j \frac{1}{j!} [(\mathbf{m}^{(k)} \pm \mathbf{w}^{(k)}) + (I \pm O)(\mathbf{u} - \mathbf{m}^{(k)})]$$

so that we obtain

$$A = I - hI + \frac{h^2}{2!}I - \cdots + (-1)^{n-1} \frac{h^{n-1}}{(n-1)!}I$$

$$B = O$$

The global error carryover is now very nearly $e^{-h}\mathbf{w}^{(k)}$, and there no longer is difficulty obtaining accurate values of the solution over wide intervals $[a, b]$.

A second instructive test problem is the differential equation $y'' + y = 0$ with initial conditions $y(0) = v_1$, $y'(0) = v_2$ or, equivalently, setting $u_1 = y$ and $u_2 = y'$,

$$\begin{bmatrix} u_1' \\ u_2' \end{bmatrix} = \begin{bmatrix} 0 & 1 \\ -1 & 0 \end{bmatrix} \begin{bmatrix} u_1 \\ u_2 \end{bmatrix}; \qquad \begin{bmatrix} u_1(0) \\ u_2(0) \end{bmatrix} = \begin{bmatrix} v_1 \\ v_2 \end{bmatrix} \tag{9.24}$$

The solution is

$$\begin{bmatrix} u_1 \\ u_2 \end{bmatrix} = \begin{bmatrix} \cos x & \sin x \\ -\sin x & \cos x \end{bmatrix} \begin{bmatrix} v_1 \\ v_2 \end{bmatrix}$$

Again taking $\mathbf{u}(x_k)$ to be $\mathbf{m}^{(k)} \pm \mathbf{w}^{(k)}$, and using degree 1 interval arithmetic, this time we find the coefficient of the $\mathbf{u}$ series term in $(x - x_k)^j$ is

$$\frac{1}{j!} \begin{bmatrix} 0 & 1 \\ -1 & 0 \end{bmatrix}^j [(\mathbf{m}^{(k)} \pm \mathbf{w}^{(k)}) + (I \pm O)(\mathbf{u} - \mathbf{m}^{(k)})]$$

If we set the matrix S equal to $\begin{bmatrix} 0 & 1 \\ -1 & 0 \end{bmatrix}$, for this problem we find

$$\mathbf{m}^{(k+1)} = \left[I + hS + \frac{h^2}{2!} S^2 + \frac{h^3}{3!} S^3 + \cdots + \frac{h^n}{n!} S^n \right] \mathbf{m}^{(k)}$$

Note that $S^2 = \begin{bmatrix} -1 & 0 \\ 0 & -1 \end{bmatrix}$ and $S^4 = I$. We find that very nearly

$$\mathbf{m}^{(k+1)} = \begin{bmatrix} \cos h & \sin h \\ -\sin h & \cos h \end{bmatrix} \mathbf{m}^{(k)} = \begin{bmatrix} m_1^{(k)}\cos h + m_2^{(k)}\sin h \\ -m_1^{(k)}\sin h + m_2^{(k)}\cos h \end{bmatrix}$$

The matrix B of 9.23 is O, and the matrix A is given by

$$A = I + hS + \frac{h^2}{2!} S^2 + \frac{h^3}{3!} S^3 + \cdots + \frac{h^{n-1}}{(n-1)!} S^{n-1}$$

The global error carryover, the $\mathbf{u}$ degree 1 wid vector $|A|\mathbf{w}^{(k)}$, is very nearly

$$\begin{bmatrix} w_1^{(k)}\cos h + w_2^{(k)}\sin h \\ w_1^{(k)}\sin h + w_2^{(k)}\cos h \end{bmatrix}$$

This is not a good result. For instance, if $w_1^{(k)} = w_2^{(k)}$, the global error carryover vector has components that approximate $(\sin h + \cos h)w_1^{(k)}$, which is close to $(h + 1)w_1^{(k)}$. After executing q of these h steps, the global error components are close to $(h + 1)^q w_1^{(k)}$. The global error grows exponentially for this problem, preventing determination of the solution $\mathbf{u}$ over a large interval $[a, b]$. Another improvement in the treatment of global error carryover is needed.

The major part of the global error carryover is generally the vector $|A|\mathbf{w}^{(k)}$. If we refer back to the derivation of the degree 1 wid in Section 7.4, we see that the term $A\mathbf{w}^{(k)}$ is converted to the form $|A|\mathbf{w}^{(k)}$ to get positive wids to combine with the $B\mathbf{w}^{(k)}$ wids. We may think of the vector $\mathbf{w}^{(k)}$ as if it were varying, its i-th component always being not greater than $w_i^{(k)}$ and not less than $-w_i^{(k)}$. Thus a wid vector defines a variation box in m space, with sides perpendicular to the coordinates. As $\mathbf{w}^{(k)}$ varies within its allotted box, the vector $A\mathbf{w}^{(k)}$ varies within an m dimensional parallelepiped. The wid vector $|A|\mathbf{w}^{(k)}$, with i-th component $\sum_k |a_{ik}|w_k^{(k)}$, defines the smallest box containing this parallelepiped.

If $\mathbf{w}^{(k)}$ were to vary within a subregion of its box, then $A\mathbf{w}^{(k)}$ would vary within the corresponding subregion image.

An improvement in global error carryover is possible by avoiding the conversion of $A\mathbf{w}^{(k)}$ to $|A|\mathbf{w}^{(k)}$. Suppose instead of maintaining the global error as a vector $\mathbf{w}^{(k)}$, we maintain it in the form $C^{(k)}\mathbf{w}^{(k)}$ using both a matrix $C^{(k)}$ and a vector $\mathbf{w}^{(k)}$. We take $C^{(1)} = I$ at the starting point $x_1 = a$. The global error carryover is now

$$(A \pm B)C^{(k)}\mathbf{w}^{(k)} = AC^{(k)}\mathbf{w}^{(k)} + BC^{(k)}\mathbf{w}^{(k)}$$
$$= AC^{(k)}\mathbf{w}^{(k)} + \mathbf{w}_B$$

where $\mathbf{w}_B = B|C^{(k)}|\mathbf{w}^{(k)}$. For the global error $C^{(k+1)}\mathbf{w}^{(k+1)}$ we take $C^{(k+1)}$ equal to $AC^{(k)}$, and $\mathbf{w}^{(k+1)}$ equal to $\mathbf{w}^{(k)}$ plus a correction vector large enough to take into account $\mathbf{w}_B$ plus the local error $\frac{h^n}{n}\hat{\mathbf{w}}_{n-1}$. If we set $\mathbf{w}' = \mathbf{w}_B + \frac{h^n}{n}\hat{\mathbf{w}}_{n-1}$, we can write

$$\mathbf{w}' = C^{(k+1)}[C^{(k+1)}]^{-1}\mathbf{w}'$$

so the correction vector $\mathbf{w}^+$ may be taken as $|[C^{(n+1)}]^{-1}|\mathbf{w}'$, and $\mathbf{w}^{(k+1)} = \mathbf{w}^{(k)} + \mathbf{w}^+$.

With the test problem 9.24 we find that the vector $\mathbf{w}^{(k)}$ now grows linearly instead of exponentially, with the addition at each h step of only a small wid vector to compensate for the local error. The matrix $C^{(k)}$ changes at each h step by being premultiplied by a matrix that is very nearly

$$\begin{bmatrix} \cos h & \sin h \\ -\sin h & \cos h \end{bmatrix}$$

and the elements of the successive $C^{(k)}$ matrices vary in the interval $[-1, 1]$. This revision of the error bounding system allows the second test problem to be followed to k decimals surprisingly large distances from the initial x value.

The system so far described works well with both test problems, but it is also possible the differential equation system is such that the increment vector $\mathbf{w}^+$ needed to adjust $\mathbf{w}^{(k+1)}$, although initially small, eventually becomes sizeable, even though the local error and the error due to the matrix B are small. An example where this occurs is the initial value problem

$$\begin{bmatrix} u_1' \\ u_2' \end{bmatrix} = \begin{bmatrix} -\frac{1}{2} & \frac{1}{2} \\ \frac{1}{2} & -\frac{1}{2} \end{bmatrix} \begin{bmatrix} u_1 \\ u_2 \end{bmatrix}; \qquad \begin{bmatrix} u_1(0) \\ u_2(0) \end{bmatrix} = \begin{bmatrix} 2 \\ 0 \end{bmatrix}$$

which has the solution

$$\begin{bmatrix} u_1(x) \\ u_2(x) \end{bmatrix} = \begin{bmatrix} 1 + e^{-x} \\ 1 - e^{-x} \end{bmatrix}$$

What occurs here is that the matrices $C^{(k)}$ approach singularity, so the elements of their inverses become ever larger, eventually making the correction vector $\mathbf{w}^+$ have huge components.

In our demonstration programs, which solve differential equations, we monitor the magnitude of the elements of the $C^{(k)}$ inverse matrix. When the largest magnitude passes a certain bound, $C^{(k)}$ is reset to I and $\mathbf{w}^{(k)}$ is reset to $|C^{(k)}|\mathbf{w}^{(k)}$.

9.7. SOLVABLE TWO-POINT BOUNDARY-VALUE PROBLEMS

To obtain a solution to the two-point boundary-value problem

$$\mathbf{u}' = \mathbf{f}(x, \mathbf{u})$$
$$\mathbf{g}(\mathbf{u}(a), \mathbf{u}(b)) = \mathbf{0}$$

$$(9.25)$$

we must find a vector $\mathbf{v}$ such that the solution $\mathbf{u}(x)$ to the initial value problem

$$\mathbf{u}' = \mathbf{f}(x, \mathbf{u})$$
$$\mathbf{u}(a) = \mathbf{v}$$

$$(9.26)$$

satisfies the boundary condition $\mathbf{g}(\mathbf{u}(a), \mathbf{u}(b)) = \mathbf{0}$. Generally a region R of m-space is prescribed in which we search for suitable vectors $\mathbf{v}$. If any of the boundary conditions is of the form $u_j(a) = c$, where c is some constant, this simplifies the search, because the j component of $\mathbf{v}$ is fixed at c, and the dimension of the search region is reduced by one. We use the notation $\mathbf{u}(x; \mathbf{v})$ to denote the solution to the initial value problem corresponding to $\mathbf{v}$, so that the boundary condition can be written as $\mathbf{g}(\mathbf{v}, \mathbf{u}(b; \mathbf{v})) = \mathbf{0}$ or, more simply, as $\mathbf{G}(\mathbf{v}) = \mathbf{0}$, where $\mathbf{G}(\mathbf{v})$ is $\mathbf{g}(\mathbf{v}, \mathbf{u}(b; \mathbf{v}))$. The two-point boundary-value problem is essentially a problem of finding zeros of the function $\mathbf{G}$. For each zero we find, the solution to the boundary problem is obtained by numerically solving the initial value problem 9.26 with $\mathbf{v}$ set equal to the zero.

We assume that the search region for $\mathbf{v}$ is a box B in $\mathbf{v}$-space, possibly of dimension less than m, if some $\mathbf{v}$ components are fixed by the boundary conditions. For such a box, we must be certain that $\mathbf{u}(x; \mathbf{v})$ is defined for $\mathbf{v}$ in B and x in $[a, b]$. So in general we need an appropriate containment box $B_{[a,b],M}$ satisfying the Lipschitz condition. We phrase the solvable problem so that the goal is to approximate the $\mathbf{G}$ zeros that lie in the search box B. If we can obtain a zero accurately enough, then we can obtain the corresponding solution of the boundary problem to the number of decimal places we require. We model the problem after the solvable problems of Chapter 5:

SOLVABLE PROBLEM 9.5: Given the two-point boundary-value problem $\mathbf{u}' = \mathbf{f}(x, \mathbf{u})$, $\mathbf{g}(\mathbf{u}(a), \mathbf{u}(b)) = \mathbf{0}$, with $\mathbf{f}$ and $\mathbf{g}$ being elementary, with the function $\mathbf{G}(\mathbf{v}) = \mathbf{g}(\mathbf{v}, \mathbf{u}(b; \mathbf{v}))$ defined for $\mathbf{v}$ in a box B, having only isolated zeros in B, and with $\mathbf{G}(\mathbf{v})$ nonzero on the boundary of B, given a containment box $B_{[a,b],M}$ satisfying the Lipschitz condition and containing all the functions $\mathbf{u}(x; \mathbf{v})$ for $\mathbf{v}$ in B, bound all the $\mathbf{G}$ zeros in B by giving to k decimals

1. points identified as simple zeros, or
2. points identified as containing within their tilde-box a region where $\max_i |G_i(\mathbf{v})| < 10^{-k}$.

In some of the Chapter 5 solvable problems, for example, Problem 5.6, a search of the boundary of the box B is specified to find points where the problem function is close to zero. And the topological degree helps to resolve whether a type 2 point actually is a zero. Although in principle these possibilities are open to us, the difficulty in obtaining a value for our problem function $\mathbf{G}(\mathbf{v})$ precludes these luxuries in a practical program, so they have not been included in the statement of Problem 9.5.

The k-decimal values of a point of type 1 or 2 can be used to generate the components of the corresponding solution $\mathbf{u}(x; \mathbf{v})$ to an accuracy that diminishes as x gets farther from a. A "simple zero" point definitely gives a solution to the boundary-value problem, while a type 2 point gives only a solution possibility. Such a point obtained for a certain k could be missing for a larger k.

The procedure of Chapter 5 for finding the zeros of a function $\mathbf{f}$ can be adapted to Problem 9.5. The procedure of Chapter 5 has two parts. The search region B is repeatedly subdivided into smaller and smaller subboxes, and intervals are found for the components of $\mathbf{f}$ over these subboxes. A subbox is discarded whenever a component interval does not contain 0. Periodically, for the sets of subboxes remaining, container boxes are constructed, each bounding off a group of mutually adjacent subboxes, and for these container boxes an attempt is made to obtain intervals for the partial derivatives of the $\mathbf{f}$ components. We need to show how the two parts of this procedure can be carried out for our function $\mathbf{G}$.

The inequality 9.13 implies that if we have intervals for the components of $\mathbf{v}$, we can obtain intervals for the components of $\mathbf{u}(b; \mathbf{v})$. These $\mathbf{u}(b; \mathbf{v})$ intervals may be overly wide, but their widths can be made to shrink toward zero as the widths of the $\mathbf{v}$ intervals shrink toward zero. Intervals for $\mathbf{u}(b; \mathbf{v})$ lead to intervals for the components of $\mathbf{G}(\mathbf{v})$. Thus there is no fundamental difficulty carrying through the first part of the Chapter 5 method.

As for the container box procedure, recall that the Chapter 5 solvable problems do not require that partial derivatives be defined for $\mathbf{f}$. When they are defined, there is a possibility of obtaining points known to be simple zeros by a Newton's method procedure. Determining that a container box contains a single simple zero of $\mathbf{G}$ requires computing intervals for the partial derivatives of $\mathbf{G}(\mathbf{v})$. This requires computing intervals for the partial derivatives of $\mathbf{u}(b;\mathbf{v})$ with respect to the nonfixed components of $\mathbf{v}$. When we differentiate the differential equations for $\mathbf{u}(x;\mathbf{v})$ with respect to these components, we obtain the equations

$$\frac{\partial u_i'(x;\mathbf{v})}{\partial v_j} = \frac{\partial f_i(x,\mathbf{u})}{\partial v_j} \qquad i,j = 1,\ldots,m$$

which also may be written as

$$\left(\frac{\partial u_i(x;\mathbf{v})}{\partial v_j}\right)' = \sum_k \frac{\partial f_i(x,\mathbf{u})}{\partial u_k}\frac{\partial u_k(x;\mathbf{v})}{\partial v_j} \qquad i,j = 1,\ldots,m \qquad (9.27)$$

If the function $\mathbf{f}(x,\mathbf{u})$ has all the required partial derivatives, then, in principle at least, because we have ordinary differential equations for the partial derivatives $\frac{\partial u_i(x;\mathbf{v})}{\partial v_j}$, we can obtain intervals for $\frac{\partial u_i(b;\mathbf{v})}{\partial v_j}$ over a $\mathbf{v}$ box just as we obtain intervals for $u_i(b;\mathbf{v})$ over a $\mathbf{v}$ box. We consider practical methods of getting such intervals next.

9.8. SOLVING THE BOUNDARY-VALUE PROBLEM BY POWER SERIES

We dispense with having to provide a containment box $B_{[a,b],M}$ by allowing an appropriate escape, and to ensure that power series methods can be used, we require the function $\mathbf{f}$ to be infinitely differentiable.

SOLVABLE PROBLEM 9.6: Given the two-point boundary-value problem $\mathbf{u}' = \mathbf{f}(x,\mathbf{u})$, $\mathbf{g}(\mathbf{u}(a),\mathbf{u}(b)) = \mathbf{0}$, with $\mathbf{f}$ and $\mathbf{g}$ being elementary, with $\mathbf{f}$ having all partial derivatives with respect to all its variables when x is in $[a,b]$, with the function $\mathbf{G}(\mathbf{v}) = \mathbf{g}(\mathbf{v},\mathbf{u}(b;\mathbf{v}))$ having only isolated zeros for $\mathbf{v}$ in a given box B, and with $\mathbf{G}(\mathbf{v})$ nonzero on the boundary of B, bound all the $\mathbf{G}$ zeros in B by giving to k decimals

1. points identified as simple zeros, or
2. points identified as containing within their tilde-box a region where $\max_i |G_i(\mathbf{v})| < 10^{-k}$,

or find to k decimals a point $\mathbf{v}_0$ in B where for some x in $[a,b]$ a component of $\mathbf{u}(x;\mathbf{v}_0)$ is larger in magnitude than a specified positive bound M.

In Sections 9.5 and 9.6, which are devoted to the initial value problem, we used series methods to obtain interval bounds for $\mathbf{u}(x)$ at various x arguments in $[a, b]$. With those methods we start with the representation $\mathbf{m}^{(k)} \pm \mathbf{w}^{(k)}$ for $\mathbf{u}$ at x_k, and then obtain a similar representation for $\mathbf{u}$ at a point x_{k+1} further to the right. Those methods can be used in the first part of the zero finding procedure for the boundary-value problem, where interval bounds are required for $\mathbf{u}(b; \mathbf{v})$ when $\mathbf{v}$ is restricted to some subbox $B^{(S)}$ with center $\mathbf{v}^{(S)}$ and wid $\mathbf{w}^{(S)}$. Here, instead of starting at $x = a$ with the $\mathbf{u}$ representation $\mathbf{v} \pm \mathbf{0}$, as we did for the initial value problem, we start with the $\mathbf{u}(a)$ representation $\mathbf{v}^{(S)} \pm \mathbf{w}^{(S)}$, and compute $\mathbf{u}$ for increasing x until we obtain an interval representation $\mathbf{m}^{(k')} \pm \mathbf{w}^{(k')}$ for $\mathbf{u}(b)$, which becomes the interval representation of $\mathbf{u}(b; \mathbf{v})$ over $B^{(S)}$. Then a degree 0 interval computation for $\mathbf{g}(\mathbf{v}, \mathbf{u}(b; \mathbf{v}))$ gives us an interval representation of $\mathbf{G}(\mathbf{v})$ over $B^{(S)}$. Thus the first part of the Chapter 5 procedure for locating zeros can be carried out.

The second part of the procedure requires that interval bounds be obtained for the elements of the Jacobian matrix for $\mathbf{G}(\mathbf{v})$ over a container box $B^{(C)}$. Here we need to do an interval degree 1 series computation for $\mathbf{g}(\mathbf{v}, \mathbf{u}(b; \mathbf{v}))$. If the mid-wid representation for the box $B^{(C)}$ is $\mathbf{v}^{(C)} \pm \mathbf{w}^{(C)}$, the degree 1 series representation for $\mathbf{v}$ over this box is

$$\mathbf{v} = (\mathbf{v}^{(C)} \pm \mathbf{w}^{(C)}) + (I \pm O)(\mathbf{v} - \mathbf{v}^{(C)})\bullet$$

We need a representation for $\mathbf{u}(b; \mathbf{v})$ of the form

$$\mathbf{u}(b; \mathbf{v}) = (\mathbf{m}^{(k')} \pm \mathbf{w}^{(k')}) + (A^{(k')} \pm B^{(k')})(\mathbf{v} - \mathbf{v}^{(C)})\bullet$$

Here we are supposing that somehow the partial derivatives of $\mathbf{u}(b; \mathbf{v})$ have been bounded in the intervals defined by matrices $A^{(k')}$ and $B^{(k')}$. With its variables $\mathbf{v}$ and $\mathbf{u}(x; \mathbf{v})$ set this way, a degree 1 series expansion of the elementary function $\mathbf{g}$ gives us the $\mathbf{G}$ partial derivative intervals we need. In our expression above for $\mathbf{v}$, we presumed that none of the $\mathbf{v}$ components had been fixed as constants by the boundary conditions. If component j of $\mathbf{v}$ is fixed at the value c, then its degree 0 term becomes $c \pm 0$, and its degree 1 terms all have coefficients 0 ± 0.

We can obtain the required partial derivative intervals for $\mathbf{u}(b; \mathbf{v})$ over a box $B^{(C)}$ by extending the procedure for obtaining intervals for $\mathbf{u}(b; \mathbf{v})$ over $B^{(S)}$. Here we not only are bounding in intervals the solution to the differential equation $\mathbf{u}' = \mathbf{f}(x, \mathbf{u})$ at selected points x_k in $[a, b]$ but must bound in intervals the solution to the differential equations 9.27. Suppose then that at a point x_k in $[a, b]$ we have obtained the interval vector $\mathbf{m}^{(k)} \pm \mathbf{w}^{(k)}$ for $\mathbf{u}(x_k; \mathbf{v})$ over the box $B^{(C)}$ and have obtained the interval array $A^{(k)} \pm B^{(k)}$ for the Jacobian elements of $\mathbf{u}(x_k; \mathbf{v})$ over the same box. The representative for $\mathbf{u}$ as a series in $\mathbf{v}$ is

$$\mathbf{u} = (\mathbf{m}^{(k)} \pm \mathbf{w}^{(k)}) + (A^{(k)} \pm B^{(k)})(\mathbf{v} - \mathbf{v}^{(C)})\bullet \qquad (9.28)$$

In Section 9.5 there were two parts to the procedure for obtaining a series for $\mathbf{u}$ in $[x_k, x_k + h]$. In the first part, a containment box for $\mathbf{u}$ is obtained, with h being determined in the process and the local error found. In the second part, the global error for $\mathbf{u}$ at x_{k+1} is found. Because we are solving differential equations just as before, our procedure has the same two parts. At the point x_k, we have $\mathbf{u}$ given by 9.28. (Initially, when $k = 1$ and $x_1 = a$, then $A^{(1)} = I$ and $B^{(1)} = O$.) In the first part of the procedure, variables are set to

$$x = (x_k \pm h) + (1 \pm 0)(x - x_k)\bullet$$

$$\mathbf{u} = (\mathbf{m}^{(k)} \pm \hat{\mathbf{w}}) + (A^{(k)} \pm \hat{B})(\mathbf{v} - \mathbf{v}^{(C)})\bullet$$

The wid vector $\hat{\mathbf{w}}$ and wid matrix $\hat{B}$ have components that are larger than or identical to corresponding components of $\mathbf{w}^{(k)}$ and $B^{(k)}$. For the chosen step width h, we determine whether the $\hat{\mathbf{w}}$ and $\hat{B}$ estimates are acceptable by the outcome of a series computation for $\mathbf{f}$ that is degree 0 in x:

$$\mathbf{f} = (\mathbf{f}_0 \pm \hat{\mathbf{w}}_0) + (A_0 \pm \hat{B}_0)(\mathbf{v} - \mathbf{v}^{(C)})$$

(In this series computation, $\mathbf{u}$ is viewed not only as a function of x but also as a function of the $\mathbf{v}$ components, which are variables independent of x. If an x degree k computation of $\mathbf{f}$ is made, the expression always is obtained to $\mathbf{v}$ degree 1. This gives us the needed x degree k values for $(\frac{\partial u_i}{\partial v_j})'$.) Now we make the containment test

$$\mathbf{w}^{(k)} + h(|\mathbf{f}_0| + \hat{\mathbf{w}}_0) < \hat{\mathbf{w}} \qquad B + h(|A_0| + \hat{B}_0) < \hat{B}$$

If any component of $\hat{\mathbf{w}}$ or $\hat{B}$ fails this test, it is replaced by the corresponding component on the other side of the inequalities above, times a factor slightly above 1, say 1.1, and the test is repeated after $\mathbf{f}$ is recalculated. After a fixed number of these tests fail, h is halved, and the whole process restarted.

After $\hat{\mathbf{w}}$ and $\hat{B}$ are found, the next step is to obtain all the higher degree x terms of the $\mathbf{u}$ series. Note that if the x degree k term of $\mathbf{f}$ is determined to be

$$[(\mathbf{f}_k \pm \mathbf{w}_k) + (A_k \pm B_k)(\mathbf{v} - \mathbf{v}^{(C)})](x - x_k)^k$$

then the x degree $k + 1$ term of $\mathbf{u}$ is

$$\left[\left(\frac{1}{k+1}\mathbf{f}_k \pm \frac{1}{k+1}\mathbf{w}_k\right) + \left(\frac{1}{k+1}A_k \pm \frac{1}{k+1}B_k\right)(\mathbf{v} - \mathbf{v}^{(C)})\right](x - x_k)^{k+1}$$

All the other parts of the power series computation described in Section 9.5 have their counterparts in this computation. The difference is that now we compute additional series terms in $\mathbf{v}$ when we make our $\mathbf{u}$ power series computations.

The method for improving wid computations described in Section 9.6 needs to be used here also to keep wid values from becoming unnecessarily large. There are more intervals now to be viewed as variables than previously, these being all the intervals appearing in 9.28.

9.9. THE LINEAR BOUNDARY-VALUE PROBLEM

The boundary-value problem becomes easier to treat if the differential equations are linear and the boundary conditions are linear. In this case the two-point boundary-value problem itself is called *linear*. The function $\mathbf{f}(x, \mathbf{u})$ has the vector–matrix representation

$$f(x, \mathbf{u}) = D(x)\,\mathbf{u} + \mathbf{h}(x)$$

where $D(x)$ is an m-square matrix whose elements are functions of x, and $\mathbf{h}(x)$ is an m-vector whose components also are functions of x. If all these functions of x are elementary and defined in $[a, b]$, then according to Solvable Problem 9.4, the components of $\mathbf{u}(b; \mathbf{v})$ can be determined for any $\mathbf{v}$.

Because the differential equations are linear, it is possible to obtain the following representation for $\mathbf{u}(b; \mathbf{v})$:

$$\mathbf{u}(b; \mathbf{v}) = Q\mathbf{v} + \mathbf{p} \tag{9.29}$$

Here Q is a real m-square matrix and $\mathbf{p}$ is a real m-vector. Column j of Q equals the solution at $x = b$ of the initial value problem.

$$\mathbf{u}' = D(x)\mathbf{u}, \qquad \mathbf{u}(a) = \mathbf{e}_j \tag{9.30}$$

with $\mathbf{e}_j$ taken equal to column j of the m-square identity matrix. The vector $\mathbf{p}$ equals the solution at $x - b$ of the initial value problem

$$\mathbf{u}' = \mathbf{h}(x), \qquad \mathbf{u}(a) = \mathbf{0} \tag{9.31}$$

If for each j we multiply by v_j the solution to 9.30 and sum, and then add the solution to (9.31), we obtain the solution to the initial value problem

$$\mathbf{u}' = D(x)\mathbf{u} + \mathbf{h}(x), \qquad \mathbf{u}(a) = \mathbf{v}$$

If we use Equation 9.29 in the linear boundary condition equation

$$A\mathbf{u}(a) + B\mathbf{u}(b) = \mathbf{c}$$

we obtain the matrix equation

$$A\mathbf{v} + BQ\mathbf{v} = \mathbf{c} - B\mathbf{p} \tag{9.32}$$

Thus if the matrix $A + BQ$ is nonsingular, there is only one vector $\mathbf{v}$ that gives a solution to the boundary-value problem, the vector $(A + BQ)^{-1}(\mathbf{c} - B\mathbf{p})$. Note that the matrix $A + BQ$ equals the Jacobian matrix of $\mathbf{G}(\mathbf{v}) = A\mathbf{v} + B\mathbf{u}(b, \mathbf{v}) - \mathbf{c}$. For the linear boundary-value problem, a possible solvable problem is

SOLVABLE PROBLEM 9.7: Given the linear two-point boundary-value problem

$$\mathbf{u}' = D(x)\mathbf{u} + \mathbf{h}(x), \qquad A\mathbf{u}(a) + B\mathbf{u}(b) = \mathbf{c}$$

where the elements of the matrix $D(x)$ and the components of the vector $\mathbf{h}(x)$ are elementary functions of x defined in $[a, b]$, find to k decimals the lone simple zero of $\mathbf{G}(\mathbf{v}) = A\mathbf{v} + B\mathbf{u}(b, \mathbf{v}) - \mathbf{c}$ or indicate that the magnitude of the $\mathbf{G}$ Jacobian is less than 10^{-k}.

This problem can be solved numerically by finding $m + 1$ separate solutions to various initial value problems, namely the solutions whose values at $x = b$ are needed to form Q and $\mathbf{p}$. But generally it is more efficient to get these values by using the procedure described in the preceding section for obtaining the $\mathbf{G}(\mathbf{v})$ partial derivatives, because then we need only one sweep through $[a, b]$ with the differential equation solution procedure.

EXERCISES

1. Call `difsys` and solve the simple initial value problem $u' = u$, $u(0) = 1$, (solution $u = e^x$) to 10 decimals for x in $[0, 1]$, spacing the points at which the solution is displayed by the amount 0.1.

2. Call `difsys` and solve the system of equations 9.8 to five decimals in the interval $[0, 5]$. Use 9.8 for the constant g and 0.1 for v_0, and take the display spacing to be 0.05.

3. Define the variable θ to be the angle a pendulum makes with the vertical axis. A free rigid weightless pendulum of length l with a mass m at its free end satisfies the equations below.

$$\frac{dx}{dt} = l \cos \theta \frac{d\theta}{dt}$$

$$\frac{dy}{dt} = l \sin \theta \frac{d\theta}{dt}$$

$$ml^2 \frac{d^2\theta}{dt^2} = -gl \sin \theta$$

Call `difsys` to solve these equations with the same initial conditions and settings as the problem of Exercise 2, and check that the obtained answers for x and y agree with the answers of Exercise 2. You need to save the file `difsys.prt` obtained for Exercise 2 as, say, `dif-sys2.prt`. Then with an editor program, you can compare that file with the `difsys.prt` file for this problem.

4. Call `difsys` and solve the initial value problem for $y'' + y = 0$ to 10 decimals in the interval $[0, 20\pi]$. Take the initial conditions to be $y(0) = 0$ and $y'(0) = 1$ (with solution $y = \sin x$), and use a display spacing of 0.1π.

5. Call `difsys` and solve to 5 decimals the initial value problem

$$u_1' = (-.5 - K)u_1 + (-.5 + K)u_2, \qquad u_1(0) = 2$$

$$u_2' = (-.5 + K)u_1 + (-.5 - K)u_2, \qquad u_2(0) = 0$$

The solution is $u_1 = e^{-x} + e^{-2Kx}, u_2 = e^{-x} - e^{-2Kx}$. Set the parameter K initially to 10, with the x display spacing taken as 0.1, and obtain the solution in the x interval $[0, 1]$. Notice that the e^{-2Kx} terms of the solution rapidly diminish in significance as x varies from 0 toward 1, and u_1 and u_2 approach each other as x increases. Edit the file `difsys.log` to change K to 100, and solve the problem once more by calling `difsys < difsys.log`. Repeat with K equal to 1000. As K increases, the `difsys` step integration width decreases and the time to generate the solution in the interval $[0, 1]$ increases. When K is large, the problem is often called *stiff*. Various definitions of a "stiff initial value problem" have been proposed, and we take a problem to be *stiff* when a Lipschitz constant l for its differential equation system must be large. Note that if the error bound given by 9.12 for an approximate solution $\mathbf{u}^{(\epsilon)}$ is accurate for our problem, then as l increases, a k decimal approximation to $\mathbf{u}$ in $[a, b]$ can be obtained only by decreasing ϵ, implying smaller integration steps in a power series solution attempt. Although there is no general way of making stiff problems tractable, often one can adapt specific problems to improve the speed of computation. Thus for our problem with $K = 10000$, we could approximate both u_1 and u_2 in $[0.01, 1]$ with their average $u_A = (u_1 + u_2)/2$ which has the simple equation $u'_A = -u_A, u_A(0) = 1$.

6. Call `difbnd` and solve the string eigenvalue problem given in Section 9.2 to 10 decimals, taking the search interval for u_3 to be $[0, 10]$. The other variables u_1 and u_2 are "fixed" and do not need search intervals. Use a display interval of $\pi/12$.

7. Call `difbnd` and attempt to solve the linear two-point boundary-value problem $y'' + y = 0$, $y(0) = y(\pi) = 0$. The function $y(x) = a \sin x$

with a arbitrary satisfies the problem, so there are an infinite number of solutions. With this problem, difbnd takes the escape of Solvable Problem 9.7. If the boundary conditions are changed to $y(0) = y(\pi/2) = 0$, or to $y(0) = 0$, $y(\pi/2) = 1$, difbnd the unique solution. Change the problem by editing the file difbnd.log and calling difbnd < difbnd.log. Solve these problems to 10 decimals, and use a display spacing of 0.1π.

8. Call difbnd and solve the two-point boundary-value problem $y'' + e^y = 0$, $y(0) = y(1) = 0$. Use a search interval for y' of $[0, 12]$ to locate two separate solutions, one with $y'(0)$ near 0.5 and one with $y'(0)$ near 11.

9. The two-point boundary-value problem $y'' = \lambda\sinh(\lambda y)$, $y(0) = 0$, $y'(1) = 1$ is often used as a difficult test case for numerical programs. For any positive λ there is a solution, but the solution gets more sensitive to the correct value of $y'(0)$ as λ increases. Call difbnd and locate to 10 decimals the solution when $\lambda = 1$, using the $y'(0)$ search interval $[0, 1]$. Then take $\lambda = 5$, and use the narrower search interval $[0, 0.05]$.

10. Call difbnd and solve to only one decimal the boundary-value problem $y' = y^2$, $y(0) = y(1)$. Take the search interval for $y(0)$ to be $[-0.5, 0.5]$. The solution $y(x) = 0$ is obtained after a long computer run because option 2 of Solvable Problem 9.6 must be taken, the faster Newton method procedure of option 1 not being applicable because computed Jacobian intervals always contain 0. (difbnd identifies option 2 solutions as possibly spurious). Repeat the problem with the differential equation changed to $y' = y^2 + 10^{-6}$. This time there is no solution, but nevertheless difbnd locates an option 2 solution. This example is similar to the findzero example of Exercise 5.5. Repeat the changed problem with the number of decimals increased from 1 to 2. This time no solution is found.

NOTES

N9.1. Three general references for the problems of this chapter are the books by Ascher, Mattheij, and Russell [1], Coddington and Levinson [3], and Hairer et al [4, 5].

N9.2. The dissertation by Lohner [6] gives an excellent analysis of the methods of calculating error when solving problems of differential equations.

REFERENCES

1. Ascher, U. M., Mattheij, R. M., and Russell, R. D., *Numerical Solution of Boundary Value Problems of Ordinary Differential Equations*, Prentice Hall, Englewood Cliffs, 1988.
2. Brenan, K. E., Campbell, S. L., and Petzold, L. R., *Numerical Solution of Initial-Value Problems of Differential-Algebraic Equations*, Elsevier, New York, 1989.
3. Coddington, E. A. and Levinson, N., *Theory of Ordinary Differential Equations*, reprint edit., Robert E. Krieger Publ. Co., Malabar, FL, 1987.
4. Hairer, E., Nørsett, S. P., and Wanner, G., *Solving Ordinary Differential Equations I, Nonstiff Problems*, Springer series in computational mathematics 8, Springer-Verlag, Berlin, 1987.
5. Hairer, E. and Wanner, G. *Solving Ordinary Differential Equations II, Stiff and Differential-Algebraic Problems*, Springer series in computational mathematics 14, Springer-Verlag, Berlin, 1991.
6. Lohner, R., *Einschliesung der Lösung gewöhnlicher Anfanfs- und Randwertaufgaben und Anwendungen*, Doctorate of Science dissertation, Faculty of Mathematics of Karlsruhe University, 1988.

X

THE C++ SYSTEM FOR PRECISE COMPUTATION

10.1. INTRODUCTION

The C++ language, a successor to the popular language C, started to come into widespread use in the later 1980s. A key feature of C++ is its "class" system for defining a new type, the type accorded the same rank in the language as predefined types such as `integer`. A C++ compiler, when it encounters a standard programming symbol for an operation with the new type, such as *, +, -, /, assigns whatever action the class designer has coded for the symbol. After a new type class is defined, and code written for all the appropriate operations, one can write programs in terms of the new type just as one can for a predefined type, that is, by using programming statements that mimic mathematical statements. In this way the language allows itself to be extended, with an ease that few other languages can match.

In the next section we list the various problem solving programs that are available, many of which have been described in earlier chapters. All of their C++ text files are contained in the CD-ROM accompanying this text, readable by a PC using any Microsoft Windows operating system. The CD-ROM also contains compiled ready-to-run versions of the demo programs, so you can test the demos without a C++ compiler. When you are ready to write your own precise computation programs, then you will need an appropriate C++ compiler program to convert your text files into executable files.

All the demo C++ text files can also be obtained via the web. The text files and a few other useful files have been combined into a single long file, the *range module*. Section 10.3 describes the procedure for downloading the range module and dissecting the module into its components. Section 10.4 has general information on how to use a compiler to convert the text files into executable files.

10.2. DEMONSTRATION PROGRAMS

Below are listed the module demo programs with a brief description of what they do. Most programs compute answers correct to the number of decimal places the user specifies, up to a preset bound. A program's upper bound on the number of correct decimal places is either to keep the display of answers intelligible or to keep the execution time of the program from becoming too long.

Name	Purpose	Relevant Text
calc	Computes quantities like a simple calculator	2.1–2.6
triangle	Solves triangles for unknown sides and angles	
findzero	Finds where n functions of n variables are simultaneously zero	5.1–5.9
equat	Solves a system of linear equations	7.2
i_equat1	Solves a system of interval linear equations for interval answers	7.3–7.5
deriv	Finds derivatives or partial derivatives of a function	4.1–4.3, 4.9
integ	Finds the definite integral of a function $f(x)$	4.4–4.7
impint	Finds the improper integral of a function $f(x)$	4.8
integ_n	Finds the n-dimensional integral of a function $f(x_1, \ldots, x_n)$	4.9, 4.10
difsys	Solves the initial value problem for a system of ordinary differential equations (ODEs)	9.1–9.6
difbnd	Solves the two-point boundary-value problem for a system of ODEs	9.7–9.9
roots	Finds the roots of a polynomial with real coefficients	6.1–6.5
eigen	Finds the eigenvalues and eigenvectors of a real square matrix	7.9–7.12
maxmin	Finds the maximum or minimum of a function of n variables	8.1
findextr	Finds where a function of n variables has a zero gradient	8.2
r_calc	Computes rational quantities like a simple	2.7

	calculator	
r_equat	Solves a system of rational linear equations for rational answers	7.13
i_equat	Solves a system of interval linear equations for interval answers	7.15
r_roots	Finds the roots of a polynomial with rational coefficients	6.1–6.5
r_eigen	Finds the eigenvalues and eigenvectors of a rational square matrix	7.13
linpro	Solves a linear programming problem	7.14
c_calc	Computes complex quantities like a simple calculator	6.6
c_deriv	Finds the derivatives of a complex analytic function	6.6
c_equat	Solves a system of complex linear equations for complex answers	7.2
c_fdzero	Finds where n complex functions of n variables are zero	6.7, 6.8
c_roots	Finds the roots of a polynomial with complex coefficients	6.1–6.5
c_eigen	Finds the eigenvalues and eigenvectors of a complex square matrix	7.9–7.12
problem	Displays a menu of problem types, and invokes the demo needed	

The demo programs that work problems which may take many keyboard entries to define, keep a record of the entries in a like-named file of extension log. Thus after completing a run with difsys you will find an entry record in difsys.log. The keyboard record is convenient when working a set of similar problems. After the first problem is solved, the log file can be edited to specify the next problem and then made standard input. In unix or MSDOS, the command to do this would be difsys < difsys.log.

A demo program that creates a log file also puts a record of a successful run in a like named file of extension prt for later viewing or printing. Thus after completing a run with difsys, the details of the run appear in the file difsys.prt.

10.3. OBTAINING THE RANGE MODULE VIA THE WEB

The module is available from the Texas A&M University Mathematics Department through the world wide web system. The starting www address to use is given below.

```
http//www.math.tamu.edu/~aberth/soft.html
```

Follow the instructions on the displayed page. Two sets of instructions are given, one for `unix` computers and one for `PC`s. In either case one obtains a module file containing all the parts of the system and then by giving the additional commands listed disects this file into its component parts. One of the component files, `describe.tex`, has general information about the demo programs in the `TeX` language.

10.4. CONVERSION OF THE DEMO TEXT PROGRAMS INTO EXECUTABLE FILES

You need an up-to-date C++ compiler to convert the module text files into object files and then to link the object files into executable files. The text files use the standard C include files `ctype.h, stdlib.h,` and `string.h`, and the standard C++ include files `iostream.h, fstream.h,` and `iomanip.h`. These files normally come with a C++ compiler system. All module files have been checked with the following C++ compilers: Free Software Foundation's GNU gcc compiler for SunOS (version 2.7.2) and Borland's C++ Borland's C++ compiler (version 5).

If your compiler system or your operating system supplies a `make` program, the conversion of the demo text files into executable files is relatively easy. Into the subdirectory in which the demo executable files are to reside, copy all the `.cc` files, all the `.h` files, and the `makefile`. With the `makefile` in place, one need only enter the command `make` to get all the executable files. However, the beginning part of the `makefile`, which defines compiling and linking commands, may require editing to conform to your compiler system.

10.5. THE C++ RANGE ARITHMETIC SYSTEM

In our description of range arithmetic in Section 2.4, we allowed the mantissa of a ranged number to change in length by a single decimal digit and used a single decimal digit for the range. We presented this type of range arithmetic because it is easy to describe and easy to understand. With our C++ system, mantissa digits are stored using the type `integer`, and it is necessary to take into account the number of decimal digits that it is convenient to store in an `integer` unit or the *decimal width* of an `integer`. The decimal width depends on the C++ compiler values for `sizeof(int)` and `sizeof(long)`, and the default value is four, a suitable value for all C++ compilers. The number of digits in a mantissa is always a multiple of the decimal width, and the number of digits in the range is the decimal width.

To revise the range arithmetic system of Section 2.4 to accommodate a decimal width of m digits, change each mantissa digit d_i and range digit r in the representation

$$(\text{sign})0.d_1d_2 \ldots d_n \pm r \cdot 10^e$$

from a single digit to a block of m decimal digits, and denote these by the symbols D_i and R. The exponent base, which was 10, is changed to $B = 10^m$, that is, 1 followed by m zeros. A ranged number now has the representation

$$\{(\text{sign})0.D_1D_2 \ldots D_n \pm R\} \cdot B^E$$

For example, with a decimal width of four, the constant 17.52 has the representation

$$\{+ 0.[0017]\ [5200] \pm [0000]\} \cdot 10000^1$$

The C++ system is essentially identical to the range arithmetic system described in Chapter 2, except that base 10 with its 10 digits 0 through 9 is replaced by a base allowing more digits, specifically a base with 10,000 different digits if the decimal width is 4.

Our C++ system has two basic classes for convenient range arithmetic: `real` and `string`. The file `real.h` defines both classes. Generally, each ranged number variable is a `real` and is converted into `string` form for console display or printing. A keyboard entry is converted into a `string`, which then can be used to set a `real`'s value.

In the following sections, we give a general description of these and other classes, so that the demo text files can be better understood. We assume the reader has had some experience with programming languages but not necessarily with the C++ language. We do need to introduce some C++ terminology here. An instance of a class is called an *object*. Thus the program statement "real X, Y;" sets up two `real` objects. (We often refer to `real` objects as "variables.") A class *member* is a function which may take any number of arguments of various types and return any type but must be associated with a class object, an implied extra argument of the function. The object name and function name are separated by a period. Thus if X is a `real` object, the program fragment X.entry(A,B) identifies entry(A,B) as a `real` member, that in this particular case is associated with X.

10.6. CLASS `real`

The examples below show the ways of initializing reals.

```
real X = 22, Y = "2.4 e2", Z = X; or
real X(22), Y("2.4 e2"), Z(X);
```

A decimal constant of any length may be assigned to a `real` if the constant is enclosed by quotes. Here the usual constant notation "e" or "E" to signal a following power of 10 may be used. A constant such as `220` that gets converted by the compiler to type `int` (integer) or `long` (long integer) does not require quotes.

Certain constant `reals` are provided, and their use saves constructor time. These are `zero`, `one`, `two`, `ten`, `tenth`, `half`, `pi` and `logten`. Thus `real X = ten` is equivalent to `real X = "10"`, or to `real X = 10`. Note that a `real` not initialized is set equal to `zero`, so that `real X` is equivalent to `real X = zero`.

For `real` computations, there are the four rational operators +, −, *, and /, a prefix minus, and the exponentiation operator ^. These are used in the same way as floating-point operators would be used. There is also a range modification operator %, where `X % E` returns `X` with its range increased by the maximum absolute value of `E`. (In earlier chapters, for this operation we used the symbol ⊕, defined in Section 2.5.)

The `real` comparison operators ==, !=, >, >=, <, and <= are equivalent to the relations $\doteq$, $\ne$, $\gtrdot$, $\geqq$, $\lessdot$, and $\leqq$, which we used in earlier chapters. A few tests besides comparisons are also provided. For instance, if `X` is a `real`, then `X.is_zero()` tests whether `X` is an exact zero, and `X.is_int()` tests whether `X` is an exact integer.

Any `real` variable defines an interval, and the midpoint and halfwidth of that interval can be obtained as exact `reals` by using the members `mid()` and `wid()`. Thus to obtain these two parameters for `X`, we write `X.mid()` and `X.wid()`.

10.7. CONTROL OF THE PRECISION OF RANGE ARITHMETIC

The precision of range arithmetic can be varied from a low of 12 decimal digits, the default value, up to almost 40,000 decimal digits. The higher precision is set, the slower range arithmetic performs, and the more memory space is needed, so it is important not to set precision excessively high. The constant `pi` (π) is used extensively in the evaluation of the `real` trigonometric functions, and the constant `logten` (ln 10) is used in logarithm calculations. Each of these constants is defined to exactly 1000 decimal places, so with any `real` computation requiring these constants, precision must be kept under the 1000 decimal digit bound.

There are several functions that access and alter the precision of range arithmetic. One of these is `set_precision(n)`, which sets the precision to n decimal digits. Actually, precision changes by multiples of the decimal width, so the setting is at least n decimal digits but possibly a few digits higher. The function `current_precision()` returns the true value, in decimal digits, of the current precision setting.

After the precision of computation has been set to its initial value, the precision of an executing program can be controlled by the `test` member and `add_precision` function. If X is a `real`, then `X.test(t, n)` leads to an examination of X to determine if it could be displayed to n correct decimals in a format determined by the integer t. If t is 1, the format is fixed decimal, e.g., `325.5573`; if t is 2, the format is scientific floating point, e.g., `5.3456 E3`. The outcome of this examination affects the global variable `test_failure`, which always equals 0 as a program begins execution. If X meets the display specification, `test_failure` is unchanged. If X fails by k digits from it, then `test_failure` is changed to k, but only if this represents an increase for `test_failure`. That is, the `test` operation never decreases `test_failure` from an earlier setting. After a series of `test` calls have been made, `test_failure` is at the maximum encountered deficit.

The function `add_precision(q)` increases the number of decimal digits of range arithmetic precision by the value of `test_failure` plus the integer argument q. The parameter `test_failure` is then reset to zero, so that another series of `test` calls can be made afterwards. If the argument q is omitted, it is presumed zero.

10.8. CLASS `string`

The examples below illustrate the ways a `string` can be initialized.

```
string A = "abc", B = A;     or     string A("abc"), B(A);
```

If a `string` is not initialized, it is set equal to `""`.

The operator + is provided for concatenation, along with the related operator +=. Thus if S is a `string`, then one can write

```
S = S + "defgh";     or     S += "defgh";
```

The two comparison operators == and != are provided for determining whether or not two strings are identical.

The operator >> is for keyboard entry from an input file or from standard input (`cin`). Thus `cin >> S` sets `string` S to the string of symbols subsequently entered at the console keyboard, the string terminated by hitting the <ENTER> key. If just the single letter "q" or "Q" is entered, this is interpreted as "quit" and causes the program to terminate. The operator << is for console display to an output file or to standard output (`cout`). For instance, `cout << S` causes the console display of `string` S.

10.9. CONVERSION OF A `real` TO A `string`

The classes we use for computing, like `real` and other classes not yet described, have provision for converting a class object to `string` form, permitting the object to be viewed at the console. The conversion is always done with the class member `str`, usually taking two `int` arguments. Suppose X is a `real`, and we want a representation to reside in `string` S. This is achieved with the statement

$$S = X.str(t, n);$$

Here if the integer t is 1, then X is converted to fixed decimal format with n decimal places. If t is 2, then X is converted to scientific floating-point format with n decimal places. In the conversion, the range of X is always considered, and if n correct decimal places are not obtainable, then the maximum number of correct decimal places is supplied instead. Occasionally an extra correct decimal place appears, which occurs (see Section 3.4) when n correct decimal places cannot be supplied, but n + 1 correct places can. If t is 3, then X is converted to integer format; in this case the n argument is unnecessary and may be omitted. If both t and n are omitted, then the `str` member makes the decision as to the best form of output and number of decimals to be displayed; the maximum number of decimals consistent with X's range will be chosen.

10.10. KEYBOARD INPUT

The demo programs obtain via standard input whatever information is needed to define the problem to be solved. Standard input is the keyboard, unless the program has been called by a command of the form `program < infile`, in which case standard input is the file `infile`. It is frequently necessary to obtain an integer via the console keyboard and to check that the integer received is within prescribed bounds. For this purpose, the function below is provided:

```
int_entry("message", I, lo_bnd, hi_bnd,
"message_lo", "message_hi");
```

This function displays `message` at the console, converts the keyboard line to an integer, and sets the `int` argument I equal to this number if the bounds are satisfied. If they are not satisfied, the appropriate message selected from `message_lo` or `message_hi` is displayed, and the entire process begins again. The function arguments `message_lo` and `message_hi` are not required and may be omitted.

Besides integers, the demo programs need various constants and elementary functions to be entered at the keyboard, and here the demo user is allowed certain standard symbols like `sin` or `tan`, or standard operators like + or *. Class real has two members to achieve this, entry and evaluate. Consider constants first. Suppose real X is to be set to a keyboard constant. This is accomplished by the statement

X.entry("message", S);

As before, message gets displayed at the console for the benefit of the program user. The second argument S is a string to hold the entered keyboard line in case a higher precision recalculation of the entered constant is needed later. The recalculation, if it is needed, is obtained with the statement

X.evaluate(S);

Next suppose an elementary function of one or more variables is to be entered. This is accomplished by a statement of the same form as previously, except that now there are two string arguments:

X.entry("message", S, SS);

Here string S records the entered keyboard line as described previously, and string SS supplies a list of allowed function arguments, with commas separating arguments. Thus if we intend to obtain a function $f(x,y)$ by keyboard entry, then SS will have been set to the string "x,y" (perhaps by a preceding statement SS = "x,y";). This form for SS ensures that the keyboard entry sin(x) * cos(y) is accepted and that the entry sin(z) is not. Thus for the entry of a function, a second string argument must appear in the entry member, and the entered keyboard line is checked for syntax but is not evaluated as occurs with a constant. The associated object X does not get changed in any way, the end result of the entry operation being to record the syntax checked function in the string S.

To evaluate a keyboard entered function, we first must construct a real array with the i-th array element set equal to the value of the i-th function variable, and then the evaluate member is called with an extra argument identifying the real array. For instance, if the real array has the name arg, and we want to evaluate the function $f(x,y)$ whose definition now resides in the string S, the statement

F.evaluate(S, arg);

deposits in real F the value of $f(x,y)$ with x taken as arg[0] and y taken as arg[1].

Most demo programs, before obtaining constants or functions, first display at the console all the allowed symbols, via the statement

real::list_tokens();

Here `list_tokens()` is a `real` member not taking an associated class object. Such members are prefixed by the class name followed by "`::`". The `entry` program decodes an entered function or constant according to the system specified by `real::list_tokens()` and not by the system used by a C++ compiler. After a successful entry of a function or constant, the needed list of evaluation operations is recorded in the associated `string` after the entered characters. These evaluation operations are used by subsequent `evaluate` calls.

The `entry` method of reading in a function can be used without the necessity of actually typing in the function at the console. The function can be designated within a program by initializing some `string` to the function, another `string` to the function's variables, and then entry is called with an empty `message`. Thus if we want to use the function $e^x \sin yz$ in our program, we can achieve this with the statements `S = "exp(x)*sin(y*z)"` and `SS = "x,y,z"`, followed by the statement `X.entry("", S, SS)`. The value of $e^x \sin yz$ would be obtained by an `evaluate` call in the way described previously.

The `entry` member uses different interpretations for the exponentiation operator, corresponding to different mathematical usages. The function x^n is defined for all x if n is a positive integer and is defined for nonzero x if n is a negative integer or zero. A function such as $x^{0.4}$ is defined if x is positive or zero but not if x is negative. All of this is the usual mathematical usage. However, the function $x^{\frac{1}{3}}$ can be defined for negative x by using the relation $(-x)^{\frac{1}{3}} = -x^{\frac{1}{3}}$. Similarly, the function $x^{\frac{2}{5}}$ can be defined for negative x by using the relation $(-x)^{\frac{2}{5}} = x^{\frac{2}{5}}$. To accommodate this second and third usage, whenever `entry` encounters the keystroke sequence `...^ (p / q)`, where p and q are integers, if p and q are both odd, the second usage is assumed, and if p is even and q is odd, the third usage is assumed. Thus the function $x^{\frac{1}{3}}$ can be specified as `x^(1/3)`, and the function $x^{\frac{2}{5}}$ can be specified as `x^(2/5)`.

Two parameters allow program control of the behavior of `evaluate`, both parameters being set to either TRUE or FALSE (integers 1 or 0). One parameter is

`real::evaluate_error_display`

With the default TRUE setting, an `evaluate` computation error leads to an error message, and then a display of the keyboard line with an indication of the point where the error occurred. With the FALSE setting, all `evaluate` error messages are blocked from console display. The second parameter is

`real::evaluate_exit_on_error`

With the default TRUE setting, a program terminates after an `evaluate` error. With the FALSE setting, an error ends `evaluate` computation but without a program termination. When the FALSE setting is used, the parameter `real::evaluate_error` should be sampled after each `evaluate` to determine from its `TRUE` or `FALSE` value whether an error occurred.

10.11. TWO SIMPLE PROGRAM EXAMPLES

Rump [3] described a simple computation that is difficult to do correctly with ordinary floating-point arithmetic: Compute

$$f = 333.75b^6 + a^2(11a^2 * b^2 - b^6 - 121b^4 - 2) + 5.5b^8 + \frac{a}{2b}$$

with

$$a = 77617.0 \text{ and } b = 33096.0$$

When Rump evaluated f using FORTRAN at various precisions on an IBM System 370 mainframe, the answers he obtained were as follows:

$$
\begin{aligned}
\text{single precision:} \quad & f = +1.172603 \ldots \\
\text{double precision:} \quad & f = +1.1726039400531 \ldots \\
\text{extended precision:} \quad & f = +1.172603940053178 \ldots
\end{aligned}
$$

Even though all three values agree to 7 decimal digits, the true value to 15 decimal places is

$$f = -0.827396059946821\tilde{}$$

and so *all* FORTRAN values are incorrect even for the f sign! Of course the variable f could have been correctly evaluated with a floating-point precision sufficiently high, but there is no easy way of determining what that precision is. This example shows that when a floating-point computation is done at two different precisions, the matching initial digits of the two answers are not necessarily correct.

The C++ program below evaluates f and in the process displays the precision of computation being used. The program needs 60 decimal digits of precision to determine f to 15 correct decimal places. Note the program loop that is used to ensure that 15 correct decimal digits are obtained.

```
#include "real.h"
int main()
{
real var[2], f;
var[0] = "77617.0"; var[1] = "33096.0";
string variables = "a,b", fs;
```

```
fs="333.75*b^6 + a^2*(11*a^2*b^2 - b^6 - 121*b^4-2) +
    5.5*b^8 + a/(2*b)";
f.entry("", fs, variables);
set_precision(20);
while (1) {
     cout << "precision = " << current_precision() <<
     endl;
     f.evaluate(fs, var);
     f.test(1,15);
     if (test_failure) add_precision(); else break;}
cout << "f = " << f.str(1,15) << endl;
return 0;
}
```

For a second programming example, suppose we wish to evaluate a function $f(x,y)$ to be entered at the keyboard to 30 correct decimals at a series of x and y arguments that also is to be entered at the keyboard. The program given below accomplishes this.

```
#include "real.h"
int main()
{
string S, T("x,y"), arg_str[2];
real arg[2], f;
real::list_tokens();
real::evaluate_exit_on_error = FALSE;
f.entry("Enter f(x,y) ", S, T);
set_precision(40);
while (1) {
     arg[0].entry("Enter x ", arg_str[0]);
     arg[1].entry("Enter y ", arg_str[1]);
     while (1) {
          f.evaluate(S, arg);
          f.test(1,30);
          if (test_failure) {
               add_precision();
               arg[0].evaluate(arg_str[0]);
               arg[1].evaluate(arg_str[1]);}
          else break;}
     if (real::evaluate_error) continue;
     cout << "f(x,y) = " << f.str(1, 30) << endl;}
}
```

Note the two program loops, the outer one to obtain the series of x and y values and the inner one to ensure that 30 correct decimals are obtained. In the inner loop, whenever more precision is needed, the x and y arguments are reevaluated at the higher precision, because they may have been entered as expressions, such as sqrt(2) or asin(.5).

10.12. CLASS exact_real

Sometimes it is convenient to compute results *exactly*, for instance, when a key branching decision depends on the comparison of two real quantities A and B, both known to have finite decimal expansions. If we can compute exact values for A and B, we always make the correct branching decision. The class exact_real is provided to make exact computations easier. For this class, before each arithmetic operation, the precision of computation is temporarily set to its highest value, and restored to its previous value after the operation. This class is derived from class real and is defined in their common header file real.h.

10.13. CLASS rational

A rational object has two reals, the numerator and the positive denominator of a rational number p/q, with all common divisors of these two numbers removed, as described in Section 2.7. This class, defined in the header file rational.h, with the code in rational.cc, has an assortment of arithmetic and comparison operators, like the class real. But unlike the reals, when a rational arithmetic operation is executed, the precision of range arithmetic is automatically set to maximum, and so rationals do not need precision control statements.

The rational class mimics the real class in the member names used for input and output. For instance, if R is a rational, then R.str() denotes a string representation of R as a pair of integers. Thus if R happened to equal $\frac{5}{9}$, then the display 5 / 9 is obtained by the statement cout << R.str();. A rational constant or function is obtained through use of the members entry and evaluate in the same way as for real constants and functions. The member

$$\text{rational::list\_tokens();}$$

displays at the console all the symbols that can be used to define a rational constant or function.

10.14. CLASS complex

A complex object is composed of two reals, defining the real part and the imaginary part of a complex number. There are a variety of arithmetic operations, and just the two comparison operators == and !=. The header file complex.h lists the operators and members for doing complex range arithmetic, and the file complex.cc has the code. This class, like the rational class, mimics the real class in the member names used for input and output. Thus if C is a complex, then C.str(t, n) converts C to string form with the integers t and n controlling how the real and complex parts are represented. A complex constant or function is obtained through use of the members entry and evaluate in the same way as for real constants and functions. The member

$$\text{complex::list\_tokens();}$$

displays at the console all the symbols that can be used to define a complex constant or function.

10.15. OTHER CLASSES

The preceding classes form the core of the range arithmetic system, and many of the demo programs use just these classes. The other classes that are used are described below with a brief discription of their application. Each class C is defined in the header file C.h, and the code for the class is given in C.cc.

The classes vector, r_vector, c_vector, and i_vector define a vector class made up of reals, rationals, complexs, and integers, respectively. Occasionally, a vector of some other type is needed, the vector having a minimal number of operations, and for such cases the template class x_vector is provided. For each vector class, there is a corresponding matrix class that is defined in the vector header file. Thus the two statements

$$\text{vector A(5); matrix C(5,6);}$$

set up a vector A of 5 real components and a matrix C of 5 rows and 6 columns of reals. If i and j are ints with nonnegative values, then A(i) and C(i,j) designate a real vector component and a real matrix element, respectively.

A number of demo programs require that the roots of certain polynomials be found with correct ranges and that multiplicities be computed. The classes poly, c_poly, and r_poly are defined to provide such operations for a real polynomial, a complex polynomial, and a rational polynomial, respectively.

In the text, some of the described procedures use a task queue to carry out their objective. The classes list and list1 often serve in this capacity. For integration demos, task queues are obtained with the class stack.

Power series computations are the means used to compute derivatives and integrals for both real and complex functions. We use a basic class series to define the elements common to any series computation. A group of term classes (like `term1`, `term2`) for `real` computations, or `c_term` classes for `complex` computations define the elements specific to a particular application.

NOTES

N10.1. A standard reference for the C++ language is the book by the language's author Stroustrup [4].

N10.2. Two other C++ systems for solving numerical problems accurately are described in the books by Hammer, Hooks, Kulisch, and Ratz [1] and Klatte, Kulisch, Wiethoff, Lawo, and Rauch [2].

REFERENCES

1. Hammer, R., Hocks, M., Kulisch, U., and Ratz, D., *C++ Tookbox for Verified Computing*, Springer-Verlag, Berlin, 1995.
2. Klatte, R., Kulisch, U., Wiethoff, A., Lawo, C., and Rauch, M., *C-XSC, a C++ Class Library for Extended Scientific Computing*, Springer-Verlag, Berlin, 1993.
3. Rump, S. M., *Algorithms for verified inclusions—theory and practice*, in *Reliability in Computing*, R. E. Moore, Ed., Academic Press, San Diego, 1988.
4. Stroustrup, B., *The C++ Programming Language*, 3rd Edit., Addison-Wesley, New York, 1997.

INDEX

ABOUT THE CD-ROM

The software on the CD-ROM is for a PC using Windows 95 or a later Microsoft operating system. The demonstration programs for precise computation on the CD-ROM are designed to be used within the MS-DOS subsystem of Windows 95.

First transfer the programs from the CD-ROM to your hard disk. The storage space needed is small, less than 6 megabytes. A few MS-DOS commands are needed to make this transfer, and these are listed in the CD-ROM file:

```
install.txt
```

If your CD-ROM reader is assigned the prefix `d:`, you can view the contents of `install.txt` by giving the MS-DOS command:

```
type d:install.txt <enter>
```

Change the `d:` prefix if your CD-ROM reader is assigned some other prefix, such as `b:` or `e:`.